Ecology *of* Streams *and* Rivers

Ecology *of* Streams *and* Rivers

Eugene Angelier
Professor Emeritus
UNIVERSITE PAUL SABATIER
TOULOUSE III
FRANCE

Translation Editor
James Munnick
UNITE DE FORMATION ET DE RECHERCHE
LANGUES VIVANTES
UNIVERSITE PAUL SABATIER
TOULOUSE III
FRANCE

CRC Press is an imprint of the
Taylor & Francis Group, an **informa** business
A SCIENCE PUBLISHERS BOOK

First published 2003 by Science Publishers, Inc.

Published 2018 by CRC Press
Taylor & Francis Group
6000 Broken Sound Parkway NW, Suite 300
Boca Raton, FL 33487-2742

ISBN-13: 978-1-57808-256-8 (pbk)
ISBN-13: 978-1-138-46864-1 (hbk)

Visit the Taylor & Francis Web site at
http://www.taylorandfrancis.com

and the CRC Press Web site at
http://www.crcpress.com

Library of Congress Cataloging-in-Publication Data

Angelier, E. (Eugène)
[Écologie des eaux courantes. English]
Ecology of streams and rivers/Eugene Angelier.
p.cm.
Includes bibliographical references (p.).
ISBN 1-57808-256-0
1. Stream ecology. I. Title.

QH541.5.S7 A5413 2003
577.6'4--dc21

2002030452

Translation of: *Écologie des eaux courantes,* Technique & Documentation, Pairs, 2000.
French edition: © Technique & Documentation, Paris, 2000

Preface

It is estimated that, every year, some 111,000 km^3 of water falls on the continents in the form of rain or snow—of which 40,000 km^3 comes from evaporation from the oceans. This volume of 40,000 km^3 returns to the oceans by way of rivers. Running waters are thus integrated in a cycle that is particularly important on the planet—the water cycle.

Like any other body in movement, running water has a kinetic energy, a power proportionate to its mass and the square of its velocity ($P = 1/2\ mV^2$). This energy is used partly to erode soils and transport and redistribute the materials of the earth's crust. Erosion, as well as the dissolution of rocks by water, has helped transform the contours of the continents over the course of geological eras. The great Precambrian, Hercynian, and Caledonian chains were formed by the most common type of erosion, that resulting from water running off slopes.

Organisms of running water live in an environment of *moving trails*. They colonize the superficial or deep horizons of the substrate (benthos) or live suspended in the water (plankton). Those that colonize the superficial horizon of the substrate are subjected to constant erosion, a drift, just like materials of the watersheds or the bed. Their strategy consists of resisting the water current and compensating the losses caused by drift. Plankton follows the transit time of water—the time it takes a mass of water to reach a confluent or its estuary. This transit time allows development of only those species that have a short cycle and high rate of multiplication, essentially algae.

The functioning of the running-water ecosystem is different from that of terrestrial, oceanic, and lacustrine environments. The cycle of materials does not follow a vertical gradient of biosynthesis at the foliage level or superficial layers of water and biodegradation on the soil or water bed. The gradient in running waters is longitudinal, from upstream to downstream; the organic matter and minerals necessary to organisms are transported and redistributed by the water current.

Organisms of running water are thus subjected to two peculiar ecological factors that make them different from all other living things of the planet: hydraulics (current, flow) and transit time. These two dominant, limiting factors determine the life of organisms of running water and their adaptive strategy. Other ecological factors, apart from temperature, often have only an incidental role in natural water courses.

Along with arable land, water courses since time immemorial are the environment most modified by humans. They are developed to allow navigation, irrigate land, prevent floods, and tap kinetic energy (hydroelectricity). In all these modes of development, it is the hydraulic factor and transit time that are modified.

Urban concentrations and industrial development have led to the use of water courses to drain away wastes. Organic biodegradable wastes and mineral salts such as nitrogen and phosphorus cause eutrophication. The extent of this phenomenon, however, depends largely on the transit time. An excess of organic materials, as well as the wastes of certain chemical compounds and metals, leads to toxic pollution.

This work is not a treatise and has no pretensions to being exhaustive. The reader will find references to several more complete and specialized works in the bibliography. The author has made a modest attempt to demonstrate the role of dominant ecological factors on the organisms of running water and the functioning of the ecosystem, as well as the consequences of human activity.

The examples are deliberately limited in number. To the extent possible, the work covers a limited number of water courses that are well known in ecological studies. This leads sometimes to repetition between chapters. However, the reader will have a more comprehensive vision of certain rivers and streams that represent various types of European water courses.

The author would like to thank all those who have allowed him to refer extensively to their work on water courses. He is also grateful to his colleagues at the Hydrobiology Laboratory of the University Paul Sabatier, together with A. Belaud (Ecole Nationale Supérieure d'Agronomie, Toulouse), H. Decamps (Centre d'Etude des Systèmes Aquatiques Continentaux), J. Giudicelli (University of Aix-Marseille III), J. Haury (ENSA, Rennes), L. Botosaneanu (University of Amsterdam), and R. Rouch (Centre National de la Recherche Scientifique, Moulis), who agreed to read certain chapters and forwarded suggestions.

Contents

Preface *v*

Chapter 1

Running water: agent of erosion, transport and redistribution of materials of the earth's crust **1**

1. The water cycle 1
2. Erosion, transport and redistribution of materials 2
   2.1. Force of moving water 2
   2.2. Load limit and competence of current 4
   2.3. Modes of material transport 5
3. Transport in solution 6
4. Chemical characteristics of continental waters 8
5. Assessment of material transport 9

Chapter 2

General characteristics of hydrographic networks **11**

1. Drainage of watersheds 11
2. Flow of water 12
   2.1. Flow in the substrate 12
   2.2. Sub-flow 13
   2.3. Flow in the plains 13
3. Water regime 15
   3.1. Nival regime 15
   3.2. Oceanic pluvial regime 15
   3.3. Mediterranean regime 17
   3.4. Watersheds with multiple regimes 17
4. Temperature of running waters 17
5. Correlations between parameters in running water 18

Chapter 3

Organisms and ecosystems of running water **21**

1. Marine and freshwater organisms 21
2. Oceanic, lacustrine, and running-water ecosystems 22
   2.1. Functioning of terrestrial, oceanic, and lacustrine ecosystems 23
   2.2. Functioning of running-water ecosystems 24
   2.3. Conclusions 25

Chapter 4

Current and benthic organisms: chronic instability of the surface horizon of the substrate **27**

1. Adaptation to the current 27
2. Drift of benthic organisms 30
   2.1. Forms of drift 30
   2.2. Floods and drift 33
   2.3. Assessment of drift 33
3. Origin of superficial benthic population 36

Chapter 5

The hyporheic environment: continuity of the substrate **39**

1. Hyporheic fauna of superficial origin 39
2. Stygobious fauna of subterranean origin 40
3. Distribution of fauna in the hyporheic environment 41
   3.1. Origin of waters 41
   3.2. Granulometry of the substrate 43
4. Origin and biogeography of stygobious fauna 46
   4.1. Stygobious fauna originating from surface waters 46
   4.2. Stygobia of marine origin 47
5. Conclusions 48

Chapter 6

Macrophytes of running waters: a substrate for algae and fauna **49**

1. Bryophytes 49
   1.1. Colonization of the substrate 49
   1.2. Population of Bryophytes 50
2. Spermatophytes with rooted plant life 51
   2.1. Colonization of the stream environment 52
   2.2. Fauna of rooted vegetation 56

Chapter 7

Life in the water trail: plankton **59**

1. Transit time and development of plankton 59
2. Modelling of phytoplankton development and seasonal successions 62
3. Conclusions 65

Chapter 8

Fish of running waters **67**

1. Swimming and the water current 67
2. Distribution of fish on a longitudinal profile 69

3. Migration of fish 71
4. Geographical distribution of fish 72
5. Conclusions 75

Chapter 9
Temperature, biological cycles and distribution of organisms 77
1. Temperature and development of organisms 77
1.1. Temperature thresholds and temperature of maximum activity 77
1.2. Lethal temperatures, limits of indefinite survival and population growth rates 80
1.3. Temperature and time of development 81
2. Biological cycles: quiescence, diapause, mono- and polyvoltinism 84
2.1. Biological cycle of species with diapause 84
2.2. Biological cycles of species with quiescence 84
2.3. Conditions of life at altitudinal limits 87
2.4. Flight periods of insects that become flying adults 87
3. Conclusions: altitudinal distribution of fauna of running waters 88
3.1. Altitudinal succession in Turbellaria 89
3.2. Altitudinal succession in the Blephariceridae of the Central Pyrenees 90
3.3. Time-space successions in the Blephariceridae of Corsica 91

Chapter 10
Light, salts and dissolved oxygen: secondary ecological factors in running water 93
1. Light and organisms in running waters 93
1.1. Light and aquatic plants 93
1.2. Light and fauna 95
2. Dissolved salts 95
2.1. Electrolytes and aquatic flora 96
2.2. Electrolytes and aquatic fauna 97
3. Dissolved oxygen and fauna 97

Chapter 11
Food webs and energy flows 101
1. Allochthonous materials and their biodegradation 101
1.1. Inputs of allochthonous materials 101
1.2. Biodegradation of allochthonous matter 102
2. Autochthonous plant production 103
2.1. Phytobenthos and phytoplankton 103

2.2. Upstream-downstream gradient of detritic and algal particulate carbon 105
3. Consumers 106
3.1. Invertebrates 106
3.2. Fish 110
4. Conclusions 115

Chapter 12

From upstream to downstream: ecological zonation of water courses **117**

1. Types of microhabitats 118
1.1. Falls and cascades 120
1.2. Rapids 120
1.3. Aprons 120
1.4. Flats 120
1.5. Muds 120
1.6. Channels 120
1.7. Lones 121
2. Upstream-downstream zonation 121
2.1. Crenal 122
2.2. Rhithral 123
2.3. Potamal 127
2.4. Illies and Botosaneanu's zonation and the concept of fluvial continuum 129
3. The alluvial plain and its zonation 130
3.1. The alluvial plain 131
3.2. Vegetation on the banks 131
3.3. Transfers between channels and the alluvial plain 133
3.4. The mobile littoral concept 134
4. Rivers with a Mediterranean hydrological regime 134
4.1. Settlements of permanent rivers 135
4.2. Temporary streams 135
4.3. Conclusions 137

Chapter 13

Ecological impacts of development of water courses **139**

1. The Lot: a river subject to multiple developments 139
1.1. Phytoplankton 142
1.2. Benthos 142
1.3. Fish 143
2. Ecological impacts of regulated flows on the rhithron: the Verdon 145
2.1. Benthos 146
2.2. Fish 147
3. Conclusions 148

Chapter 14
From eutrophication to trophic pollution **151**
1. Eutrophication in running waters 152
 1.1. Eutrophication of the Upper Aveyron 152
 1.2. Eutrophication of the Lot 153
 1.3. Eutrophication of the Charente 155
 1.4. Eutrophication of the Vire 157
 1.5. Conclusions 159
2. Trophic pollution 161
 2.1. Processes of biodegradation 161
 2.2. Self-purification in the rhithral 164
 2.3. Self-purification in the potamal 165
3. Eutrophication and trophic pollution: two sides of the same problem 169

Chapter 15
Toxic pollution **171**
1. Outline and Definitions 171
2. Toxic pollutants 171
 2.1. Toxic organic pollution 172
 2.2. Saline pollution 174
 2.3. Chemical pollution 174
 2.4. Cumulative effect micropollutants 175
 2.5. Acidification of waters 177
3. Multiple pollution 178
 3.1. The Riou-Mort 178
 3.2. Conclusions 181

Chapter 16
Biological methods of evaluating pollution **183**
1. Methods using biochemistry or ecotoxicology 183
2. Biocoenotic methods 184
 2.1. Comparative analysis of communities 184
 2.2. Methods based on the vicariance of species belonging to a single group 185
 2.3. Methods based on a combination of benthic macro-invertebrates 187
3. Conclusions 189

Chapter 17
Conclusions 195
Bibliography *201*
Index *211*

1

Running Water: Agent of Erosion, Transport and Redistribution of Materials of the Earth's Crust

The term *continental waters* refers to all the water located within the limit of the continents. Some of it is surface water—running (streams and rivers) or stagnant (lakes and ponds). Some of it is underground and results from the infiltration of surface waters. Finally, in solid form, either snow or ice, it is estimated that there are about 30 million km^3 of water stored in the Arctic, Antarctic, and high mountain regions.

Water does not remain constantly in the same state or in the same place. Its movement is called the *water cycle* and is the result of continuous transfers between the various reservoirs that make up the hydrosphere: oceans, atmosphere, lakes, glaciers, and porous rocks.

1. The water cycle

The distillation of sea water, caused by solar radiation, is the original source of the water on the continents. It occurs even at low temperatures, given the high tension of water vapour.

The volume evaporated each year from the oceans amounts to a layer about 1 m thick. This value is obviously an average between the hot regions (more than 1.5 m in the tropics) and the temperate and cold regions (35 mm between the 80° and 90° parallels). It represents an annual volume of 425,000 km^3, of which more than 90% (385,000 km^3) returns to the ocean in the form of rain and 40,000 km^3 falls on the continents.

On the continents there is also the phenomenon of evapotranspiration—directly from soil (evaporation) and indirectly from plant life (transpiration). It is estimated at 71,000 km^3 a year.

The volume of 40,000 km^3 from the oceans and 71,000 km^3 by terrestrial evaporation and transpiration results in a total of 111,000 km^3 of rain and snow that falls each year on the continents.

Close to two thirds of this water evaporates again, mostly during the hot season. A volume of 40,000 km^3 returns to the oceans each year, either directly by way of rivers or after having contributed to the charging of underground water tables.

The water cycle on the continents ultimately involves only a small proportion of the enormous mass of water contained in the hydrosphere (about 0.8×10^{-4}). But its consequences are highly significant.

At 20°C, the change from liquid state to gaseous state necessitates an input of 585 calories per gram of water, which is released into the atmosphere during vapour condensation. Large masses of water circulate in the atmosphere, with a notable transfer from the tropical zones towards the medium latitudes. The calories consumed in the hot regions during evaporation are restored in the colder regions and thus help to render habitable lands that would otherwise be too hot or too cold.

2. Erosion, transport and redistribution of materials

Water moves with a velocity that is a function of the slope. The slope-current relationship is estimated by Manning's equation:

$$V\ (m/s) = \frac{Rh^{0.66} \times S^{0.5}}{n}$$

where V is the velocity of the current (m/s), Rh the hydraulic effect, S the slope, and n the index of the roughness of the bed.

The hydraulic effect is the ratio between the area of the submerged section of the bed and its perimeter. It is inversely correlated to the mean depth.

The index of roughness can be estimated experimentally in a channel but is more difficult to estimate in a natural water course. If it is not taken into account, it can be said that the velocity of the current is proportionate, roughly speaking, to the square root of the product *hydraulic effect* × *slope*. For a given slope and flow, the velocity of the current is as high as the depth is low.

Water running off the land collects in primary and secondary streams that bring the precipitation to rivers, in a vast hydrographic drainage network by which soil materials are also carried. Running water constitutes a vast system of erosion, transport, and redistribution of materials of the earth's crust, in solid and dissolved form. It thus contributes to transforming the relief of the continents.

2.1. Force of moving water

The force of moving water (Pb)—its kinetic energy—is proportionate to its mass (m) and to the square of its velocity (V):

$$Pb = \frac{1}{2} mV^2$$

The mass of water that flows per unit of time—the flow—is equal to the area of the submerged section of the bed (S) multiplied by the velocity:

$$m = SV, \text{ from which}$$

$$Pb = \frac{1}{2} SV \times V^2 = \frac{1}{2} SV^3$$

The force of moving water is proportionate to the cube of the velocity. Part of this force is absorbed to overcome certain resistances to movement. If there were no resistance, the velocity of water would accelerate indefinitely. The resistance is due to the viscosity of the water, the roughness of the bed (particularly important in mountain streams and rivers), and above all to turbulence phenomena.

In a river, trails of water should in principle move along the axis of the bed and parallel to the surface. In fact, the kinetic energy of water and the inertia of the bed disturb the parallelism of liquid trails. The velocity of the current is higher close to the surface and in the middle of the river than it is near the bottom and near the banks (Fig. 1). The result is turbulence phenomena that absorb a significant fraction of the force of the water. The trajectories of liquid trails are not parallel to the axis of the bed. Their velocity varies at each point in size and direction, with no regularity. The turbulence can be compared to vertical pulsations of a very short period, which lead to a continuous turnover of the water.

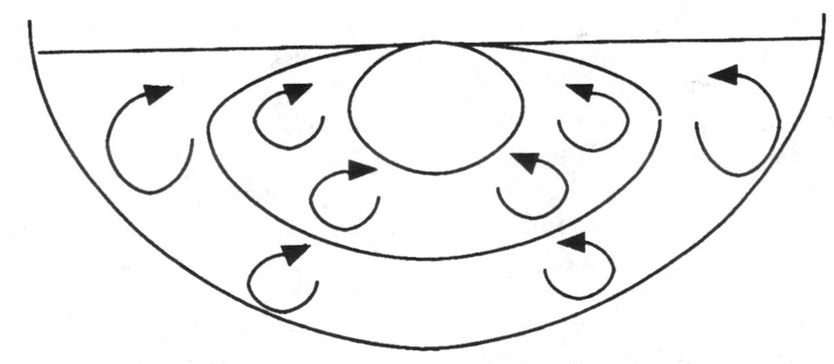

Fig. 1. Turbulence phenomena on the submerged section of a water course

From experiments in channels, a mathematical expression has been developed for turbulence, the Reynolds number:

$$Re = \frac{V \times Rh}{v}$$

where V is the mean velocity in the submerged section (in m/s), v is the coefficient of viscosity (in stock: –0.01 for water at 20°C), and Rh is the hydraulic effect.

For a stream in which the hydraulic effect is 1 m, the flow is always turbulent beyond 20 cm/s (Re > 2000) and always laminar below 5 cm/s

(Re < 500). Between 5 and 20 cm/s, turbulence is possible as a function of the roughness of the bed. The laminar flow is exceptional.

Friction and turbulence absorb most of the brute force of water. The materials transported also absorb some force. The net force of running water (Pn) is defined as its brute force less the losses due to friction, turbulence, and transport of materials:

$$Pn = Pb - (\text{friction} + \text{turbulence} + \text{material transported})$$

On the basis of the notion of force associated with slope, two types of superficial water can be defined:

— Running water has a high brute force resulting from the slope of the surrounding basin.
— Stagnant water has a weak force, resulting only from internal currents and low turbulence due to the effect of wind or temperature variations.

On a steep slope, the net force is positive. The force of the river hollows out its bed and erodes its banks, so that the load of transported materials increases, whereas the net force diminishes. When the net force is nil, the river transports its load in suspension but no longer erodes. Under these conditions, if the slope and thus the speed of the current diminish, the net force becomes negative. The river then deposits materials until its net force returns to zero. The simple action of filling in a depression suffices moreover to raise the bed and increase the slope.

At each point along the river—by simple actions of erosion, transport, and deposit of materials—what is called an *equilibrium slope* is established. It corresponds to a nil net force. The slope at each point being dependent on that of preceding and following points, the entire water course has an *equilibrium profile* (Fig. 2A), a longitudinal profile that can be compared, roughly speaking, to a branch of an asymptotic hyperbola at the base (confluent level for a tributary and sea level for a river). It is not a definitive profile. It continues to become lower since the water course continues to transport materials and thus to erode at least its upper course. The reality is less simple, because the tectonics and the nature of the land crossed play a role in the establishment of a longitudinal profile. The profile appears as a succession of thresholds and depressions determined by the hardness of rocks crossed (Fig. 2B).

2.2. Load limit and competence of current

Two concepts are important in the transport of materials: that of *load limit* and that of *competence* of the current. The load limit is the maximum weight of materials that a water course can transport as a function of its velocity, and for which the net force is nil. Competence corresponds to the maximum calibre of materials that can be transported in suspension. It is approximately proportionate to power 6 of the velocity, but this

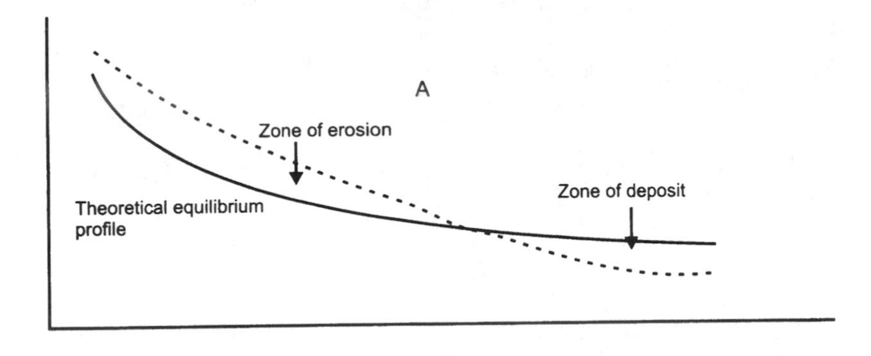

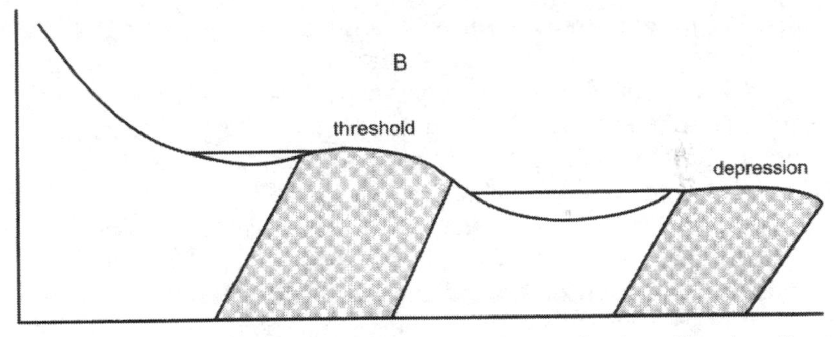

Fig. 2. Theoretical equilibrium profile and real longitudinal profile of a tributary

correlation is valid only for materials coarser than 5 mm and is itself a subject of controversy.

From the load limit and competence it can more easily be understood how materials are deposited on the profile along a river. When the slope diminishes, from upstream to downstream, the load limit decreases and materials remain on the bed. Coarser and heavier materials are deposited first (reduction of competence). The phenomenon can easily be observed in gravel deposited along a river. From upstream to downstream, the mean diameter of materials extracted from the bed becomes progressively smaller. The granulometry of the substrate is a function of the velocity of the current.

2.3. Modes of material transport

Besides being transported in suspension, the bedload may be rolled along the bed of a stream (Fig. 3). For a given velocity, the turbulence suffices to keep in suspension a particular weight of coarse materials that is a function of current competence. Coarse materials such as blocks and large pebbles are carried along by successive jumps. Turbulent ascending currents reduce the pressure of these materials on the bed, and they can thus be pushed or rolled over a few metres. Another ascending current will move and roll them anew.

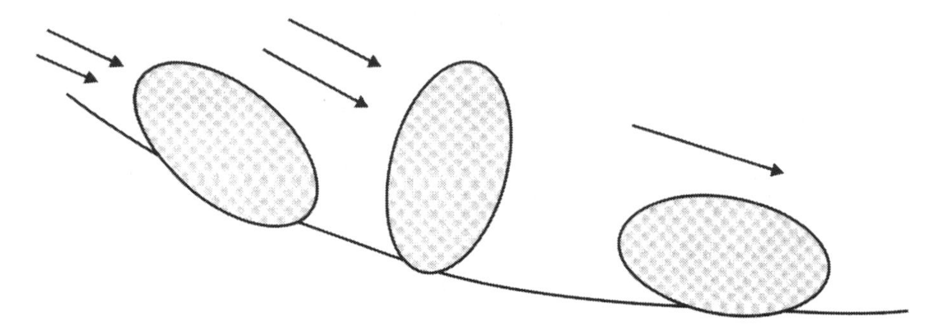

Fig. 3. Materials being rolled along the bed

The velocities of current required to shake, lift, and carry the bedload materials or to keep them in suspension are obviously different, but researchers do not always agree on their respective values. The significance of turbulence phenomena are indeed difficult to evaluate, whereas turbulence is fundamental in the movement of materials of the substrate and their maintenance in suspension.

According to Schaffernak, a current of 8 cm/s suffices to shake a 2 mm grain of sand, but 15 cm/s is necessary to move it and 11 cm/s to keep it in suspension. A current of 1 m/s shakes a pebble of 7 cm diameter, but close to 2 m/s is required to move it and 1.3 m/s to keep it in suspension. Such velocities signify that the larger pebbles are essentially rolled or moved during rises in water levels.

The adherence of materials to the bed is greater as the materials become finer, for diameters less than 0.25 mm. Consequently such materials require high current velocities to be moved along the bed, while very slow currents are sufficient to keep them in suspension. This is expressed by Hjülstrom's curves (Fig. 4).

3. Transport in solution

Electrolytes in solution in water reflect the geological nature of the land it has crossed. Granitic rocks, for example, are made up of silica, pure or in the form of silico-aluminates. Silica is poorly soluble in water in cold or temperate climates. The silicates, by alteration, yield metallic hydroxides and colloidal clays. Waters crossing granitic rocks are poorly mineralized, with a pH that is acidic or close to neutral.

Sandstone is made up of sandy grains linked by a cement. If the cement is siliceous, it is practically impervious to water. On the other hand, a carbonate or chalky sulphate cement is soluble.

Carbonic anhydride (CO_2) with water forms carbonic acid. This acid dissolves chalky rocks into soluble bicarbonates:

$$H_2CO_3 + CO_3Ca \rightarrow Ca(HCO_3)_2$$

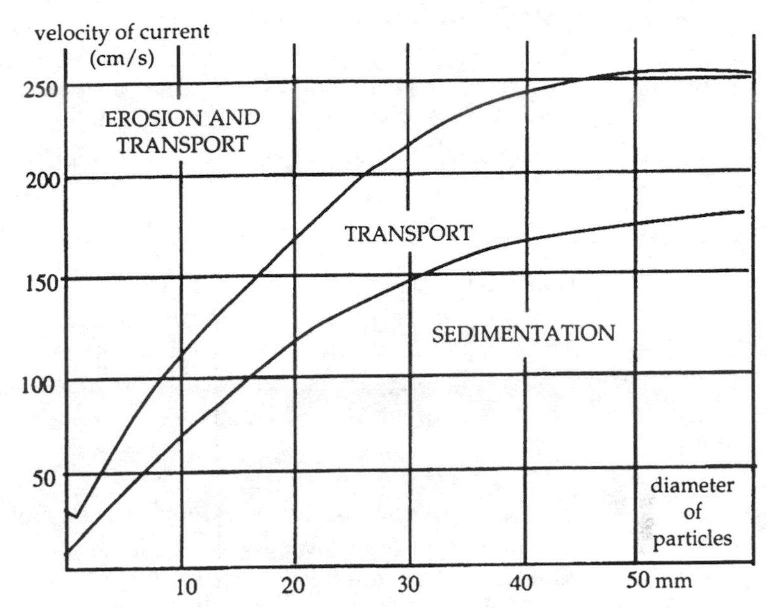

Fig. 4. Hjülstrom's curves: relations between the velocity of the current, erosion, transport in suspension, and sedimentation of materials

Waters crossing chalky terrain therefore have a high concentration of bicarbonates and a high pH.

Two rivers of the Massif Central in France—the Lot and the Truyère—clearly illustrate the influence of the terrain crossed on the electrolytes in solution (Table 1). When the Lot arrives at its confluence with the Truyère, at Entraygues, it has run along granitic and crystallophyllian terrain on its right bank and the limestone plateaux of the Causses on its left bank.

Table 1. Chemical characteristics of the waters of the Lot and the Truyère before their confluence at Entraygues, August 1971 (J.-M. Bordes, J. Cl. Luccheta and M. Rochard, 1973)

	Lot	Truyère
pH	8.5	6.5
Resistivity ($\Omega cm^2/cm$)	6665	20,236
Alkalinity (mg/l CO_3Ca)	74.8	18
Calcium (mg/l)	27.2	4.4
Magnesium (mg/l)	2	1.2
Iron (µg/l)	260	357
Sodium (mg/l)	3.4	2.9
Potassium (mg/l)	1.16	0.8
Silica (mg/l)	7.3	10.5
NO_3 (µg/l of N)	400	310
PO_4 (µg/l of P)	16	5

The Truyère has crossed practically only granites, basalts, and some crystallophyllian formations. The concentration of bicarbonates, calcium, and magnesium in the waters of the Lot and their conductivity are nearly four times those of the waters of the Truyère. The latter, however, are richer in silica. The pH is 8.5 in the Lot and only 6.5 in the Truyère.

4. Chemical characteristics of continental waters

Apart from dissolved atmospheric gases, the chemical components of water, running as well as stagnant, are electrolytes represented by anions: essentially bicarbonates, sulphates, nitrates, phosphates, silicates, and chlorides. The cations are alkaline-earths (calcium and magnesium), alkalines such as sodium and potassium, and trace metals, notably iron and manganese.

All the electrolytes do not come from the mother rock, despite the size of the latter. The nitrogenous compounds—ammoniac, nitrites, and nitrates—result from the oxidation of organic compounds on the watersheds and their synthesis from atmospheric nitrogen during storms.

Phosphates, rare in the mineral world, are equally rare in water. The use of phosphate fertilizers and polyphosphate detergents has recently helped enrich continental waters in phosphorus and contributed to the phenomenon of *eutrophication* (Chapter 14).

A significant property of natural waters is their buffering effect, due to a certain number of equilibrium reactions between CO_2, H_2CO_3, H^+, CO_3^-, HCO_3^-, Ca^{++}, and Mg^{++}:

$$H_2O + CO_2 \rightarrow H_2CO_3 \leftrightarrow H^+ + HCO_3^-$$
$$CaCO_3 + H_2CO_3 \rightarrow Ca(HCO_3)^2 \leftrightarrow Ca^{++} + 2HCO_3^-$$

The bicarbonates constitute a buffering system that stabilizes the pH of water, but they remain in this form only in the presence of free carbonic acid, called equilibrium CO_2. If this CO_2 is used up by plants during photosynthesis and disappears, part of the bicarbonates precipitates into carbonates:

$$Ca(HCO_3)^2 \rightarrow CaCO_3 + H_2O + CO_2$$

Inversely, if CO_2 is dissolved in greater quantity than equilibrium CO_2, it may solubilize carbonates into bicarbonates or even yield acid water (aggressive CO_2), if the terrain crossed is poorly carbonated.

Carbonic anhydride is therefore present in water in four different forms:

- combined in the form of neutral salt;
- combined in the form of acid salt;
- free, non-reactive (equilibrium CO_2); and
- free, reactive (aggressive CO_2).

At each point along a water course, the content of electrolytes resulting from the mother rock is correlated negatively to the flow. This is because the dissolution is a function of the area of substrate-water contact and transit time. When the flow increases (rise in water levels), the velocity of the current increases and transit time decreases, while the dissolved electrolytes are diluted in a larger volume of water.

5. Assessment of material transport

Various researchers have made annual assessments of bedload material transported by running water. For transport of suspended sediments, the most recent assessments evaluate the annual mass released into the oceans at between 13.5 billion and 15 billion tonnes, very unequally distributed among the continents.

The Yellow River deposits 850 million tonnes of material each year into the China Sea (18 kg/m^3 on average). In France, the Loire carries to Nantes only a million tonnes of material in suspension (40 g/m^3). The Nile transports 3.5 kg/m^3 of materials as it crosses Sudan, but hardly more than 1.5 kg/m^3 as it crosses Egypt. These are in reality only orders of magnitude of the mass of materials released into the oceans. Moreover, the assessments do not take into account materials deposited in the large reservoirs (estimated at a billion tonnes per year) or materials deposited in lakes, flood zones, and piedmont areas, where 50% to 90% of eroded materials are deposited along the basins.

As for the transport of suspended material, it brings each year around 2.4 billion tonnes of material, of variable nature depending on the regions. Most of the silica is released by rivers of tropical regions, but the sedimentary rocks provide most of the dissolved materials. The Loire, with a mean dry residue of 200 mg/l, releases 5 million tonnes of electrolytes into the Atlantic Ocean each year.

An erosion of 1 mm of soil corresponds to close to 2500 t/km^2/year of materials. The erosion in the Swiss Alps is about 1 m every 1000 years at 2500 m altitude. Under these conditions, within 4 to 5 million years, the Alps would be reduced to the state of modest hills if their high levels were not maintained by other material-raising phenomena. The great Precambrian and Hercynian chains were levelled by the most common type of erosion—that resulting from rainfall on the continents.

The consequence of these transfers is the existence of regions exposed to erosion, to permanent geochemical losses: these are steep mountainous regions. They are bordered by areas in which materials accumulate—the foothills of mountain chains, plateaux, inland and coastal plains—to form rich and fertile soils. When the climatic factors of temperature and humidity are favourable, it is in these deposit areas that the optimal conditions for the development of life are to be found. The Nile, for example, brings 15 t/ha of minerals and organic matter extracted from its high valleys to its floodplains in Egypt each year.

2

General Characteristics of Hydrographic Networks

1. Drainage of watersheds

Watersheds are drained by runoff water that collects in rivers and streams. The intensity of drainage can be evaluated by Horton's method, modified by Strahler (Fig. 5).

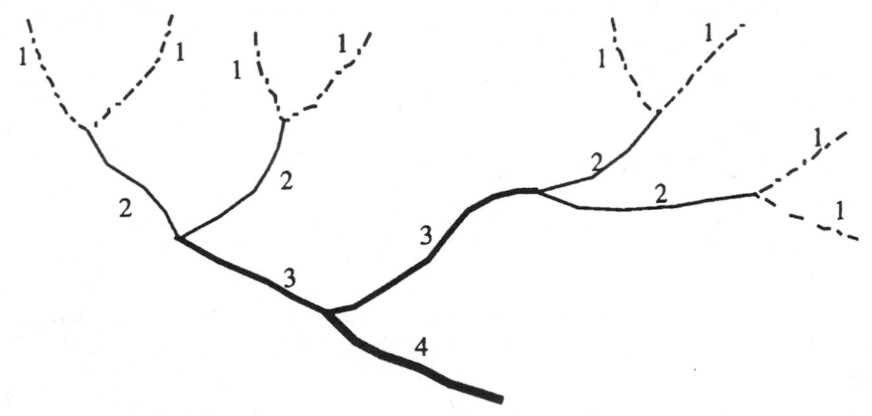

Fig. 5. Drainage order of main and secondary tributaries according to Strahler's method

Each branch of a water course can be assigned an order number. A stream that has no tributary is said to be of order 1. When two first order streams join together, they constitute a stream of order 2. From upstream to downstream, a single stream changes its order number each time it receives a stream of the same order number as its own.

The numbers of drainage order are obviously valid in cartography on a defined scale: there are more streams of order 1 on a map of 1/25,000 scale than on a map of scale 1/80,000, and the sequence of order numbers also differs.

Streams of order 1 have a particular interest: the number of such streams in a watershed reflects the intensity of drainage, expressed by the ratio of number of streams of order 1 to the area of the watershed (km^2).

The overall coefficient of drainage of the watershed of the river Lot in France, for example, is 0.074. It may be higher than 0.1 in the sub-basins of the high valley, but it is between 0.055 and 0.072 in the middle and lower valley. The area of the watershed upstream of the confluence of the Lot and Dourdou rivers (6367 km^2) represents more than half the total area, and the rainfall is also higher than it is downstream. The water regime of the Lot and the nature of the materials transported in suspension and in solution are essentially a function of precipitation and drainage of water on the upper basin.

2. Flow of water

By the process of erosion, transport, and redistribution of material, the flow of water results in a longitudinal profile of a water course that approaches an equilibrium profile (Chapter 1.2.1). The current controls a process of regularization that runs from the torrent hurtling from basin to basin to the river moving in its own alluvium. Zones of the water course moving over steep slopes and having a low order of drainage are called *rhithral*. Those moving over shallow slopes and having a deep bed, with a high order of drainage (lower course), are called *potamal*. A more precise zonation of water courses is given in Chapter 12.

2.1. Flow in the substrate

Turbulence phenomena are caused by different velocities of the current at various points of the submerged section. At the substrate level, the flow conditions are modified: the current slows and becomes practically nil. Thus, a *limit layer* is formed, which is laminar and no longer turbulent. It is thin, about one millimetre, and it is an inverse function of the velocity of the current.

Obstacles such as pebbles and blocks create areas of dead water, set apart from the main current. The water trail pulls away and moves over obstacles (Fig. 6). Downstream of the obstacle, a whirlpool forms, which moves more slowly than the main current. When the main current has a

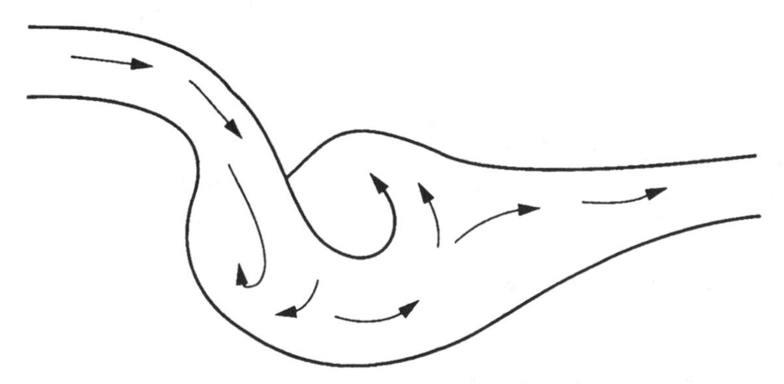

Fig. 6. Formation of whirlpool downstream of an obstacle

low velocity, the exchanges between the current and the whirlpool are low. The exchanges are greater at high velocities, while turbulence appears in the whirlpool: it is expressed in abrupt variations of pressure in the substrate.

The limit layer and zones of dead water form shelters for organisms. These shelters become unreliable at high current velocities, when turbulence or pressure variations appear on the bottom.

2.2. Sub-flow

When tributary beds are formed from their own alluvium, the visible surface flow is accompanied by a slower sub-flow. Water circulates in these alluvia at a velocity that depends on the slope and even more so on the diameter of interstices and their clogging up.

Water of the sub-flow is different from surface water in physico-chemical terms: it has lower concentrations of dissolved oxygen and certain ions, but higher overall mineralization. The surface water that penetrates the sub-flow carries to it organic matter that is a source of nutrients for the fauna that colonize that environment.

2.3. Flow in the plains

The water course in the mountain appears as an outlet of its watershed, with no other connection to it except erosion of the soil. Steep slopes favour linear flow and give it a straight track along the profile. Rivers such as the Loire, the Rhone, or the Garonne in France still have such tracks in their middle course (slopes of 0.5 to 0.6 m/km).

In the plains, water courses, especially the larger rivers, cannot be considered simple channels of water flow. When they pass rapidly from a steep area to a much flatter area, they deposit materials. There is no longer really a main bed, but a more or less complex network of channels skirting around islands or gravel banks (braided beds, Fig. 7). Rivers such as the Loire in its lower valley are straight during high water and branch between banks of gravel and sand during low water.

Meanders form on shallow slopes (Seine, 0.10 m/km from Paris to Rouen). They are cut into the alluvia or the rock itself. Meanders originate from secondary currents, mixing the waters from one bank to the other, further eroding the concave bank and depositing materials on the convex bank. Sections of tributary streams with meanders correspond to a state of equilibrium between the net force of the water and the resistance of the rock.

The curves of meanders tend to become accentuated, the concave bank hollowing out more and more: two neighbouring meanders may thus become joined. Thus, what is left of the former meander becomes isolated, a dead arm. It then evolves towards a lacustrine ecosystem, as with the lones of the Rhone upstream of Lyon.

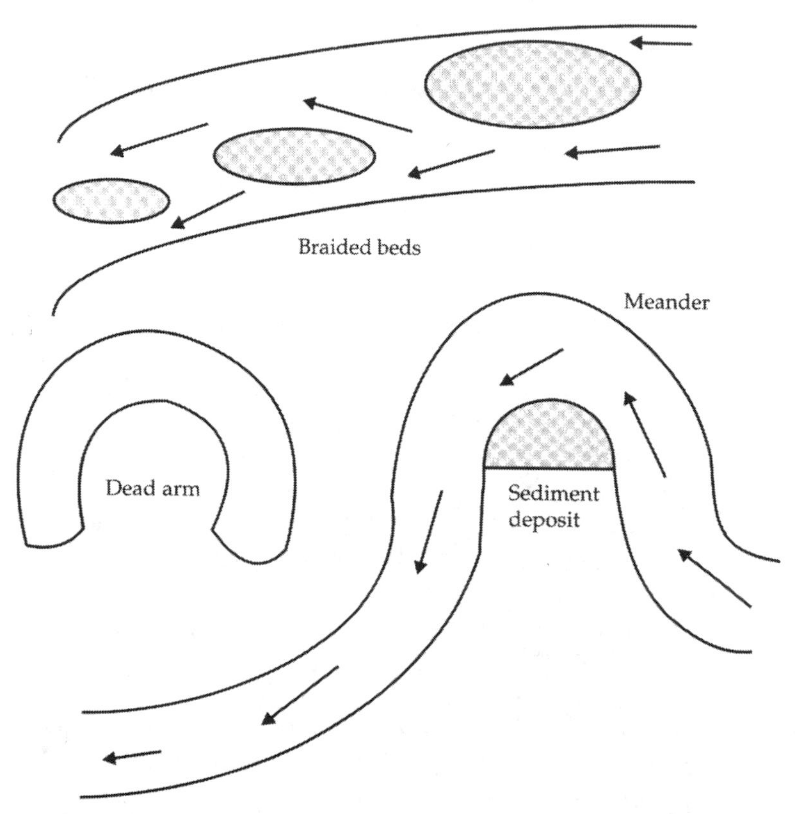

Fig. 7. Braided beds and meanders in water courses of the plains

In the alluvial plains, the weak hemming in of banks favours overflows: the ordinary bed is joined by a major bed—or floodplain—that is submerged during floods. Thus, a hydrosystem develops, its existence linked to the river's hydrodynamics. The hydrosystem is a set of aquatic, semi-aquatic, and terrestrial ecosystems, soils colonized by alluvial forests of willows, alders, and poplars (Chapter 12.3). There are equally close interactions between the water course and the water table: the Rhine, for example, contributes one third of the total ground water of the Alsace plain.

In industrialized countries and those with a dense riverside population, as in Europe, river beds have progressively become canalized and dyked in order to recover cultivable land, control floods, and facilitate navigation. Today, only fragments of the primitive alluvial forest remain. The bed of the Seine, in Paris, was twice as wide during the Middle Ages as it is now. But several river basins in the world—such as the Amazon and the Orinoco—demonstrate the role of forests subject to flooding in the hydrosystem.

3. Water regime

The flow of water courses varies throughout the year and presents alternating high water (flooding) and low water. The water regime is linked to that of the rain and the seasonal thermal cycle.

In tropical and subtropical regions, high water coincides with summer rains. In the equatorial region, two periods of high water are observed, coinciding with the post-equinox double maximum of rains.

At medium latitudes, the hydrology of the river is determined by the seasonal thermal alternation, with two basic seasons, winter and summer, and two intermediate seasons, spring and autumn. This alternation leads to three hydrological regimes—nival, oceanic pluvial, and Mediterranean—with variants depending on the altitude of the watershed.

3.1. Nival regime

On watersheds in high mountains, and at lower altitudes with a continental climate, the winter precipitation falls in the form of snow, and its flow is deferred till the spring.

In low altitudes (continental Eurasia, a large part of Canada), the snow melt is rapid in the spring—it takes place over a few weeks—and water rise is relatively brief. The low water period extends from the end of spring to the end of winter.

In the mountains, the duration of snow cover varies with the altitude. In the Alps in the Savoy region, it is around 3 months at 1200 m, 6 months at 1800 m, 8 months at 2200 m, 10 months at 2800 m, and nearly permanent at 3000 m. On a single watershed, the snow takes a longer time to melt than in a continental plains climate. The low water period starts during the summer and lasts till the beginning of spring (Fig. 8).

If glaciers occupy a significant part of the watershed, as in a polar or subpolar climate, high water persists during the entire summer (glacier regime). Conversely, if a significant part of the watershed is located below 2000 m, the first autumnal precipitation falls in the form of rain, and the spring high water is accompanied by an autumn high water, of varying importance (transitional nival regime).

3.2. Oceanic pluvial regime

Winter precipitation in the oceanic pluvial regime falls essentially in the form of rain, and it flows immediately. Evaporation from the soil and evapotranspiration being very low in winter, the coefficient of flow (ratio of flow volume to precipitated volume) is high, and the period of high water lasts from the end of autumn to the end of winter. Low water, from spring to autumn, is obviously due to less abundant rainfall, but also to intense evaporation and transpiration: the flow coefficient is reduced.

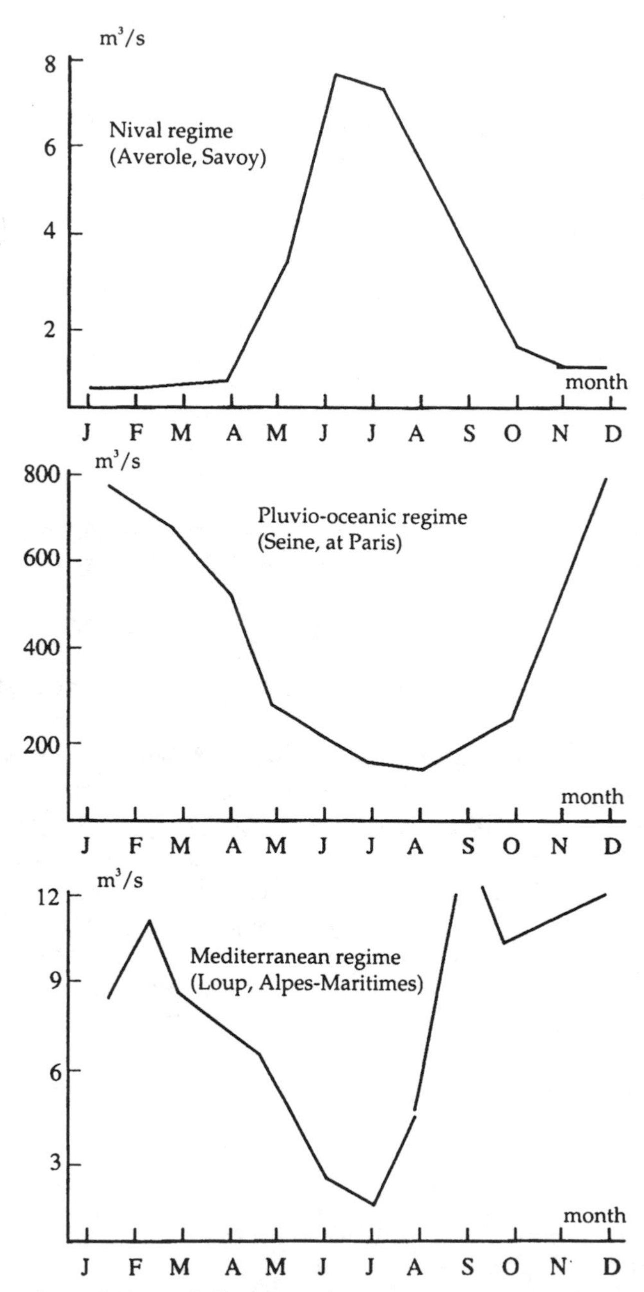

Fig. 8. Annual evolution of flow in tributaries with nival, pluvio-oceanic, and Mediterranean hydrological regimes

There is a wide range of intermediate regimes between the strictly nival and oceanic pluvial regimes, depending on the respective prevalence of mountain and plains zones on the watershed. In western Europe, the hydrographic networks, of which mountain zones are not higher than 1200 to 1500 m, have only a small nival input in high water. The regime is thus called pluvio-nival. If this input is greater, the regime is then called nivo-pluvial.

3.3. Mediterranean regime

The winter flow is significant, with two peaks, in autumn and in spring (the autumn maximum is dominant in the French rivers of the Mediterranean coast). But low summer rains, combined with intense evaporation and transpiration, lead to severe conditions of low water in summer, and even temporary drying out.

3.4. Watersheds with multiple regimes

When a river receives large tributaries with different water regimes, the regimes balance out one another. For example, the hydrological regime of the Rhone upstream of Lake Leman is glacial and nival, with a minimum flow in January and maximum flow in July, in a ratio of 1 to 6. At Lyon, the Saone river contributes its winter high water. Downstream, on the left bank, the flow of the Isère river culminates in June and that of the Durance river in May. The flow of the Rhone passes from a maximum in July in Switzerland to a maximum in January downstream of Lyon, and in June downstream of the confluence with the Isère river. At Beaucaire, the mean monthly flow varies only in a ratio of 1 to 3 between August and February, with higher flows than the annual average from December to May.

4. Temperature of running waters

In the terrestrial environment, temperature is a function of altitude, latitude, season, and insolation. The same applies to a lake environment. Other factors intervene in flowing waters: temperature of the source, distance between source and the point at which temperature is measured, and transit time of the water between these two points. The transit time depends on the slope and the flow.

When heat is studied as an ecological factor, the instantaneous measurement of temperature is insufficient; the annual thermal sum—or number of degree-days—must be taken into account (Chapter 9). This is the annual sum of daily temperatures, which indicates the thermal regime at each point of the longitudinal profile. Apart from the latitude and the altitude, the hydrological regime plays an important role in the thermal regime.

In a pluvial oceanic regime, the low water period coincides with fair weather: it starts with the spring warming and lasts until autumn. In a Mediterranean regime, the severe low water period of summer corresponds to the hottest season, and the number of degree-days is very high.

In a nival regime, the high water resulting from snow melt slows down the warming of waters until summer: the low water period lasts from summer to spring and coincides partly with the cold season. For water courses with similar summer temperatures, the number of degree-days is smaller in a nival regime than in a pluvial regime.

5. Correlations between parameters in running water

The notion of equilibrium profile such as has been defined suggests that hydraulic, morphometric, physical, and chemical parameters taken together have values that vary regularly from upstream to downstream, and that these parameters are consequently correlated.

A hierarchical classification of parameters measured on some rivers of the central Pyrenees and the Gascony hills will make clear the levels of relationships between them (Fig. 9). Distance from source to location, width of the bed, altitude, area of the watershed, and order of drainage all reflect the position of each point on a longitudinal profile. Slope, current, substrate granulometry, temperature, alkalinity, conductivity, and pH are the ecological factors that may intervene in the distribution of organisms; their values increase or decrease regularly along the longitudinal profile.

A canonical analysis on two sets of parameters, topographic and ecological (Fig. 10), indicates their correlation. The first link between the two sets is due essentially to the slope and the velocity of the current, and incidentally to the distance between source and station and to the temperature. A second link is due to the altitude of the source and, at each point of a longitudinal profile, to its altitude, the area of the watershed located upstream, and the temperature.

Two ecological factors emerge particularly from this analysis: velocity of current (linked to the slope) and temperature. The current—a hydraulic factor—and temperature are the two essential ecological factors that determine the possibilities of existence of organisms as a function of their tolerance limits. These are known as limiting factors.

The current intervenes directly as an ecological factor by its eroding action. The strategy of benthic organisms is to maintain themselves on the bottom despite the current, which tends to drag them downstream. It also intervenes indirectly. As an agent of erosion, transport, and redistribution of materials, it is the organizer of the substrate. For organisms living in suspension in water (plankton), transit time—the time water takes to reach a confluent or its estuary—plays a major role. However, it must be noted that in tropical regions, where the annual amplitudes of temperature are low, temperature does not play the essential role that it plays in higher latitudes.

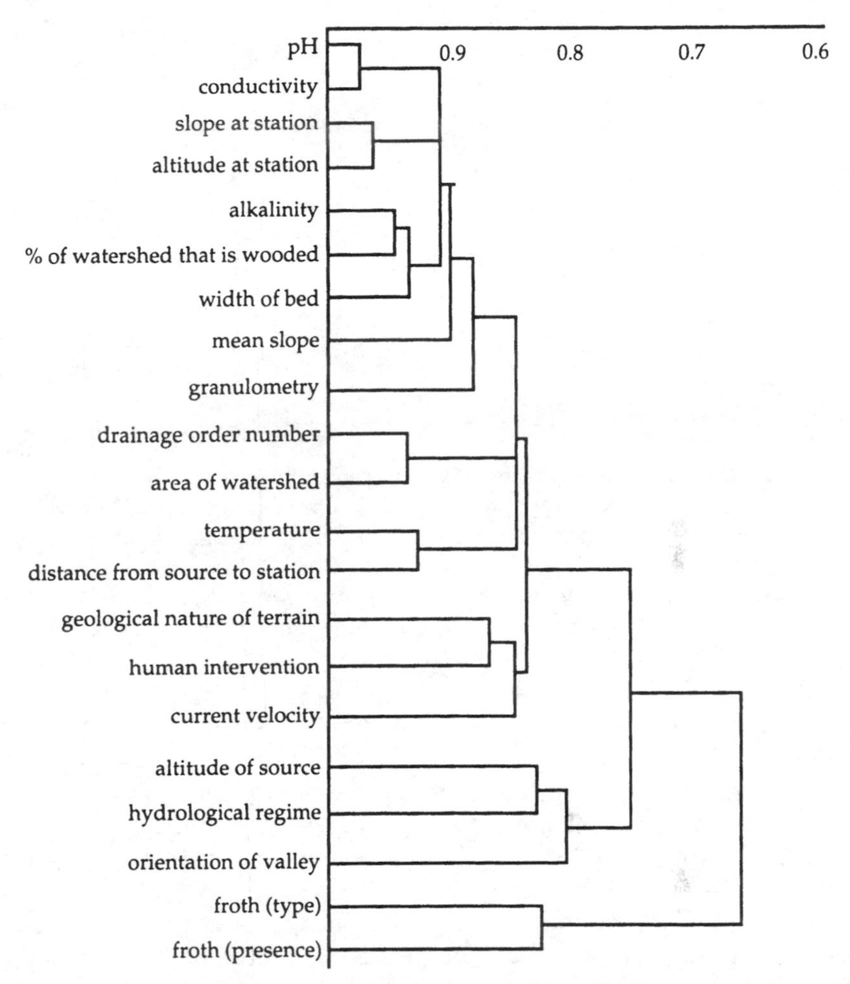

Fig. 9. Hierarchical classification of parameters of some tributaries of the central Pyrenees and Gascony hills. Correlations between the parameters evolve in the same way from upstream to downstream (E. and M.L. Angelier, J. Lauga, 1985).

Canonical analysis does not allow the introduction of qualitative parameters such as the hydrological regime. That would be possible using a factorial analysis of correspondences, which shows that the entire set of morphometric, topographic, and hydrological parameters figuring in the hierarchical classification can be represented at each point of the longitudinal profile by just five parameters:

— altitude of the source;
— altitude of the experimental station;
— slope at the experimental station;
— area of the watershed upstream of the station (EM) (or width of bed, distance between source and station, or order of drainage)

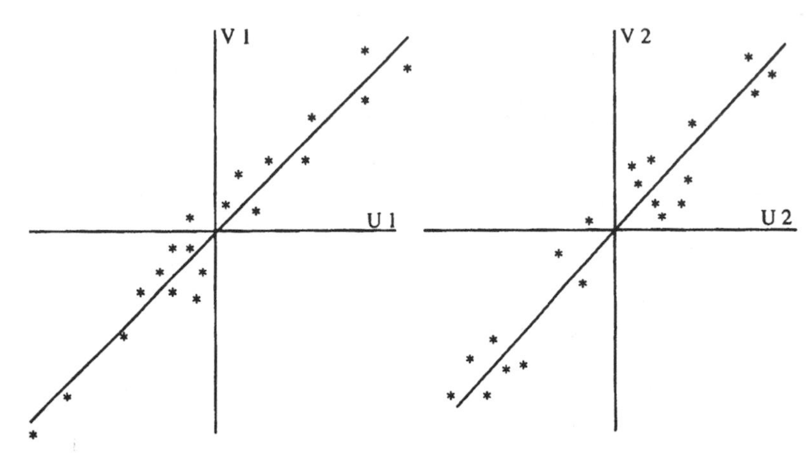

Fig. 10. Correlations between two sets of parameters, topographic and ecological, by canonical analysis. The stations (*) lie on an axis corresponding to the equation U 1 = 0.92 (slope) + 0.24 (distance between source and experimental station) and V 1 = 0.86 (velocity of current) + 0.18 (richness of fauna) + 0.32 (temperature). When the action of factors U 1 and V 1 is eliminated, a second equation appears: U 2 = 0.56 (altitude of station) – 0.55 (area of watershed) + 0.47 (altitude of source) and V 2 = 0.80 (temperature) (E. and M.L. Angelier, J. Lauga, 1985).

— hydrological regime (or mean summer temperature if the regime is poorly defined).

Thus, each point on the longitudinal profile of a watershed can be characterized by a parametric complex. It comprises a minimum of morphometric, topographic, and hydrological parameters that determine the two limiting ecological factors: current and temperature.

A canonical analysis shows that, in the central Pyrenees and Gascony hills, the different points of the longitudinal profiles lie approximately on a straight line in relation to the axes of the equations. The analysis describes an upstream-downstream continuum of the experimental station's parameters and ecological parameters. This view must, however, be tempered: at the confluence of two water courses of the same order of drainage, the flow, the area of the watershed upstream, and the width of the bed increase abruptly (theoretically doubling). Thus, the upstream-downstream continuum in fact comprises nodes. Moreover, Illies and Botosaneanu's ecological zonation is based on the presence of these nodes (Chapter 12).

3

Organisms and Ecosystems of Running Water

Oceanic and continental waters differ in their mineralization, the former being characterized by their concentration of sodium chloride. Marine and freshwater organisms constitute two isolated worlds. The number of species of marine or continental origin that colonize brackish water is limited.

1. Marine and freshwater organisms

If the major groups are considered, animals and plants, the population of continental waters is indisputably less diversified than that of the oceans. Four branches of Invertebrates (15 classes) are exclusively marine: Echinodermata, Stomochordata, Tunicata, and Cephalochordata. There are similarly 13 other classes including, among the Mollusca, the Aplacophora, Polyplacophora, Monoplacophora, and Scaphopoda, and two subclasses of Crustaceans—Cirripedia and Leptostracea.

Out of more than 20,000 species of fish, 8500 live in fresh water. But half the Invertebrates of continental waters are insects, which are practically absent in the oceans. Among the plants, Bryophytes and Pteridophytes are found exclusively in fresh water.

If species richness is referred to rather than the major groups, species richness is generally greater in continental waters. This is the consequence of the geographical isolation of continents, watersheds, and lakes, as well as the diversity of environments. All these factors taken together favour speciation. For just the aquatic Invertebrates of Europe, more than 14,000 species are presently known.

Within the population of continental waters, there are differences between populations of running and stagnant waters. Diptera Blephariceridae and Simuliidae (entirely), Ephemeroptera (90% of species), Plecoptera, and Trichoptera are among the most important groups of insects that exclusively or essentially colonize running waters.

On the other hand, Crustaceans are dominant in stagnant water, as are the Oligochaeta, Mollusca, Diptera Chironomidae, Coleoptera Dytiscidae, Heptera, and Odonata. Out of 1400 species of Chironomida of Europe, only 23% colonize running waters.

Similarities between stagnant and running environments are found at two levels:

— limnocrene sources (with very slow current) and the littoral zone of high altitude lakes;
— banks along the lower courses of rivers (potamal) and the littoral zone of low altitude lakes.

These environments are colonized by a set of species living on the moving sediments, with macrophyte vegetation—association of Oligochaeta, Mollusca, Crustacea (Copepoda, Cladocera, and Ostracoda), Diptera Chironomidae, Ceratopogonidae, Odonata, Coleoptera Dytiscidae, and Hemiptera.

Fish are most often ubiquitous. *Salmo trutta*, the common trout, for example, colonizes rivers as well as high altitude or deep lakes.

The origin of populations living in running waters depends ultimately on the upper and middle part of the water course (rhithral). In the low valleys, in the potamal, the development of a plankton drifting with the trails of water and the appearance of slow current zones or stretches of dead water constitute a transition between running and stagnant water.

The benthic flora and fauna at the superficial level of the substrate are more or less subjected to the eroding action of the current. They live on an unstable substrate—especially during a rise in the water level. The Bryophytes, fixed on the mother rock or on large blocks, conversely constitute a permanently stable environment. Bryophyte organisms are better protected from the current than benthic organisms, in that their small size allows them to survive in the sheltered environment between the dense stems and moss leaves. Sub-flows also constitute a stable environment, except during exceptionally high water periods. As for plankton, it drifts with the water; its composition and development are linked to the transit time of the water.

The conditions of existence of organisms are consequently very different depending on the environments they colonize: unstable at the water-sediment interface, stable in the Bryophytes and the sub-flow, and limited as a function of transit time for the plankton living suspended in the water and drifting with it.

2. Oceanic, lacustrine, and running-water ecosystems

An ecosystem—a set of organisms living in a delimited area as well as the resources needed to keep them alive—is characterized by a cycle of biosynthesis and biodegradation of organic matter. This cycle corresponds to the materials that pass alternately from a mineral state, oxidized, to an organic state—a state in which they flow through the ecosystem, along the food webs. The original energy necessary for organic synthesis is solar energy, transformed into chemical energy by chlorophyll function.

The functioning of terrestrial, oceanic, and lacustrine ecosystems is rather similar, while that of ecosystems of running water is very different and entirely original.

2.1. Functioning of terrestrial, oceanic, and lacustrine ecosystems

Terrestrial, oceanic, and lacustrine ecosystems are characterized by a vertical gradient of the biosynthesis-biodegradation cycle. In a forest, the leaves on trees receive sunlight and effect their own synthesis from minerals. Plants are used directly or at second hand by animals, which are primary or secondary consumers.

Everything that dies falls on to the soil, where the organic matter, the necromass, is slowly degraded, oxidized, and returned to a mineral state. The root system of a plant recovers mineral salts in soluble form from the soil.

Oceanic and lacustrine ecosystems present the same biosynthesis-biodegradation vertical gradient (Fig. 11). The biosynthesis occurs in the upper layers of the water, in relation to the depth of light penetration. The necromass accumulates on the bottom. The difference between these ecosystems and terrestrial ecosystems lies in the cycling of mineral salts.

Apart from the coastal belt of vegetation, which is an extension of the terrestrial vegetation and, like it, has a root system, it is the algae in suspension, phytoplankton, that transform solar energy into chemical energy. They are consumed by zooplankton. The necromass sinks to the bottom, is partly consumed by a detritivorous zoobenthos (analogous to soil fauna), and is mineralized. But the phytoplankton has no means of its

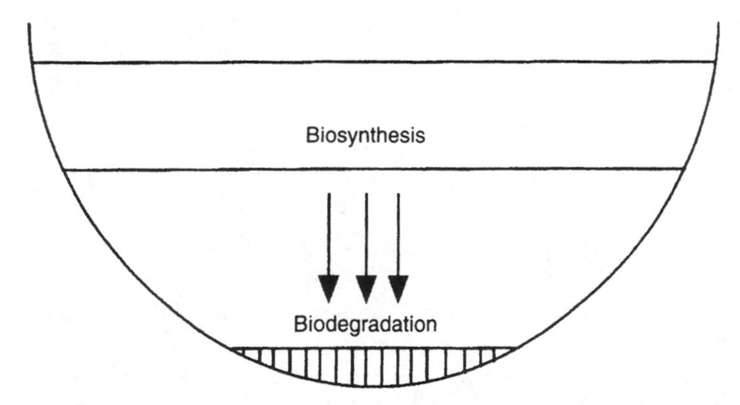

Fig. 11. Functioning of lacustrine and oceanic ecosystems. The biosynthesis-biodegradation cycle follows a vertical gradient.

own to cycle the minerals that accumulate on the bottom. This cycling is made possible only by the vertical movement of water, which carries the minerals towards the surface. Such a mode of cycling does not have the efficiency of a root system, and thus the productivity of oceanic and lacustrine waters is much lower than that of terrestrial ecosystems.

2.2. Functioning of running-water ecosystems

A water course drains a watershed and tends to reach the lowest point. The force of the moving water simultaneously erodes the materials of the watershed—resources necessary for the functioning of the ecosystem, mineral salts, and necromass—and drags living organisms with it (drift).

Running waters constitute a system of transport from upstream to downstream, and the functioning of their ecosystems follows a horizontal gradient, no longer a vertical one (Fig. 12). It is this peculiarity that makes running-water or lotic ecosystems unique. It explains their tendency to heterotrophism. Photosynthesis plays an important role only on the shallow slopes, in the plains (potamal).

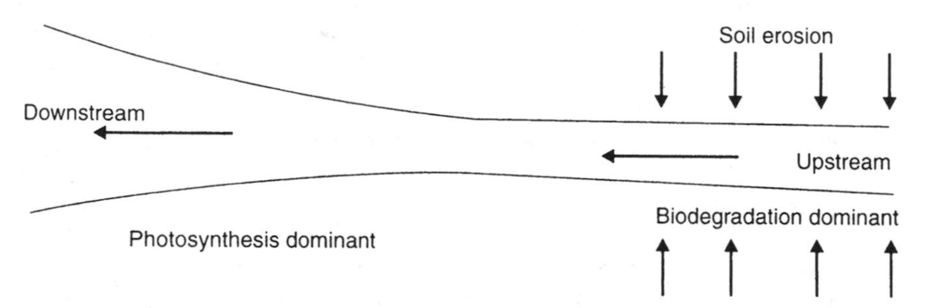

Fig. 12. Functioning of running-water ecosystems. The biodegradation-biosynthesis cycle follows a longitudinal gradient, upstream to downstream.

The horizontal functioning of lotic ecosystems is based primarily on the utilization of organic matter transported by detritivorous organisms that feed on it, on the fungi and bacteria that degrade and oxidize it. Secondly, the oxidized materials make possible the photosynthesis of fixed algae (phytobenthos) and algae in suspension (phytoplankton). The animal food webs in the upper part of the river are comparable to those of soils and are based on the necromass. It is only when the current diminishes that the phytobenthos develops, and food webs based on herbivores appear.

In oceanic and lacustrine ecosystems, exchanges of oxygen and carbon dioxide at the water-atmosphere interface are under the dominant control of photosynthesis and the respiration of organisms. This mode of control is secondary in running waters, where turbulence phenomena are dominant and more easily balance the gaseous exchanges between water and atmosphere. The appearance of a vertical biosynthesis-biodegradation gradient is an exceptional and temporary summer phenomenon in the potamal, mostly when embankments raise the water level to allow navigation.

The mode of functioning of lotic ecosystems is thus very different from that of oceanic and lacustrine ecosystems, because of the horizontal gradient along which energy and matter flow. Benthic organisms use this flow only partly, in passing. Planktonic organisms drift with it but use it optimally only when there is enough time—the water transit time—available for their development.

Contrary to what is observed in other ecosystems, the flow of energy and matter in running water appears more like a backdrop for the ecosystem than a regulator of communities.

In the hierarchy of ecological factors, the current—hydraulics—figures at the top, followed by temperature. The regulation of communities is dominated by physical factors and, in this sense, the ecosystems of running water are thought of as eternally pioneer ecosystems.

2.3. Conclusions

Limnology was originally defined as the oceanography of lakes. Lakes were studied as reduced oceans, using identical concepts. It was A. Thienemann, in 1925, who extended the field of limnology to the body of continental waters, looking at factors that influence life in fresh water.

The very peculiar functioning of running-water ecosystems brings us back to the original definition of limnology. The study of running water is based on concepts other than those of lacustrine waters—and especially on the fundamental role of hydraulics, the primary limiting factor. The horizontal gradient of energy and matter flow, the organizing role of current on the substrate and communities, and the problems of stability or instability of different substrates make running water a unique chapter in the study of continental waters.

4

Current and Benthic Organisms: Chronic Instability of the Surface Horizon of the Substrate

1. Adaptation to the current

Any organism living in flowing water—a rheophile—has mechanisms adapted to resist the current. These mechanisms are morphological and behavioural. The tendency to face the current and move against it is quite common in rheophilous animals, as is their dynamic morphology. The most common morphological characteristics are a hydrodynamic form and dorsoventral flattening. For animals living in open water, the form presenting the least resistance to the current is that of a hydrodynamic body in which the largest transversal section is located at a little more than one third of the total length. This is what is observed in many types of fish. Dorsoventral flattening appears more as an adaptation to avoid the current in benthic invertebrates. They are small, adhere to the substrate, and are protected in the boundary layer. They thus escape the pull of the current. Examples are Turbellaria, Ancylidae Mollusca, or even larvae of Ephemeroptera such as Heptageniidae (Fig. 13, Table 2).

Insect larvae cling tightly to the substrate by temporary fixations: abdominal suckers (Diptera Blephariceridae) or even a disc in the form of a sucker surrounded by a crown of hooks on the posterior part of the body (Diptera Simuliidae). In the Ephemeroptera *Rhitrogena*, the first wide pair of gills forms a sort of sucker that improves adherence to the substrate, apart from the powerful tarsal claws. The larvae of Trichoptera without a sheath moor themselves on the substrate by means of two hooks located on the last abdominal segment. As for Trichoptera with sheaths, the sheath may be weighted by stones (Goeridae) or have hydrodynamic extensions (Thremmatidae). These help anchor the organisms to the bottom. Benthic algae attach themselves on the substrate by a mucus, and Bryophytes use their rhizoids or suckers to attach themselves.

Resistance to the current obviously depends on the mode of fixation or attachment to the substrate. Only Blephariceridae and Simuliidae larvae,

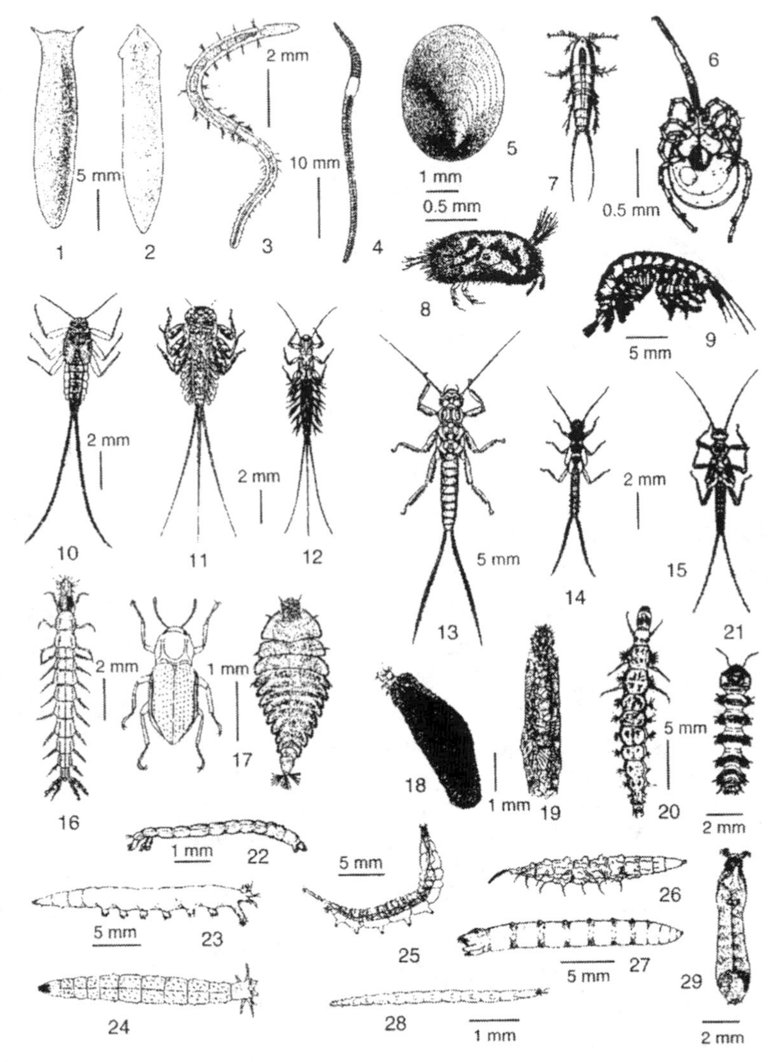

Fig. 13. Types of invertebrates characteristic of running waters. 1. *Polycelis felina*, Turbellaria. 2. *Dugesia gonocephala*, Turbellaria. 3. *Nais variabilis*, Oligochaeta. 4. *Eiseniella*, Oligochaeta. 5. *Ancylus fluviatilis*, Mollusca. 6. *Pseudotorrenticola rhynchota*, Hydrachnellae. 7. *Canthocamptus*, Copepod Harpacticide. 8. *Herpetocypris*, Ostracoda. 9. *Gammarus pulex*, Amphipoda. 10. *Baetis fluminum*, Ephemeroptera. 11. *Ecdyonurus*, Ephemeroptera. 12. *Habroleptophlebia*, Ephemeroptera. 13. *Perlodes*, Plecoptera. 14. *Leuctra*, Plecoptera. 15. *Nemoura*, Plecoptera. 16. *Orectochilus*, Coleoptera. 17. *Elmis* larva and adult, Coleoptera. 18. *Hydroptila*, Trichoptera. 19. *Pticolepus*, Trichoptera. 20. *Rhyacophila*, Trichoptera. 21. *Liponeura*, Diptera Blephariceridae. 22. Chironominae, Diptera. 23. Hemerodrominae, Diptera. 24. Tipulidae, Diptera. 25. Limnobiidae, Diptera. 26. Athericidae, Diptera. 27. Rhagionidae, Diptera. 28. Ceratopogonidae, Diptera. 29. *Simulium*, Diptera Simuliidae (Angelier, 1950; Brocher, 1913; Pattee, 1981; Richoux, 1982; Tachet et al., 1987).

which have suckers, can resist currents faster than 2 m/s. They are dominant on the steepest slopes. *Ancylus fluviatilis* (Mollusca), *Dugesia gonocephala* (Turbellaria), and *Rhitrogena semicolorata* (Ephemeroptera) hold on in the face of currents close to 1 m/s. Species such as *Ecdyonurus venosus* (Ephemeroptera) or *Isoperla oxylepis* (Plecoptera), however, do not resist currents faster than 0.60 m/s. All these invertebrates are characteristic of the rhithral.

The range between the maximum speed tolerated experimentally and the speed tolerated in natural conditions is small for the species fixed on the substrate. However, it must be remembered that current-meters cannot be used to evaluate speed of current in the boundary layer, where it is much slower. Species not fixed to the substrate can temporarily resist rapid currents, but they habitually position themselves in areas sheltered from the current—in areas of still water, under rocks and pebbles, or in interstices in the substrate. The adaptation is more behavioural than morphological, and the problem is one of energy: to limit the energy needed to avoid being carried away by the current. Although common trout and barbel resist currents of 4.4 and 2.4 m/s respectively, they usually live in areas of still water or crevices in the banks during rest periods. The crustaceans Amphipodes Gammaridae look for a shelter between pebbles on the substrate. When they swim, they avoid the major current. If they temporarily find themselves in a rapid current, they move by creeping, using their antenna and exploiting any available shelter to protect themselves. The behaviour of Turbellaria recalls that of the Gammaridae: there is a possibility that they can immobilize themselves in calm areas near the major current, adhere to the substrate, and choose a travel path that reduces the risks of being dragged along. The resistance of Turbellaria to being dragged along depends moreover on how rheophilous the species is—it is greater in *Crenobia alpina* and *Dugesia gonocephala* than in *Polycelis nigra* and *P. felina*. This last species significantly prefers currents

Table 2. Maximum velocity of current at which some invertebrate species remain fixed (Dittmar, 1955)

	Max. velocity (m/s)
Liponeura cinarescens, Diptera	< 3
Simulium sp., Diptera	2.8
Rhyacophila sp., Trichoptera	1.22
Ancylus fluviatilis, Mollusca	1.18
Rhitrogena semicolorata, Ephemeroptera	0.96
Dugesia gonocephala, Turbellaria	0.93
Baetis sp., Ephemeroptera	0.84
Isoperla oxylepis, Plecoptera	0.60
Ecdyonurus venosus, Ephemeroptera	0.57
Radix ovata, Mollusca	0.48

slower than 0.20 m/s. Another way of avoiding the current is through ambulatory behaviour—by moving between the interstices of the substrate.

Indeed, apart from some fixed forms, most benthic invertebrates live sheltered from the current. When they are in contact with it, temporarily or permanently, the organisms face the risk of being swept away and transported along with the sediments.

2. Drift of benthic organisms

The term *living drift* refers to the carrying and transport of organisms by the current. It is a phenomenon analogous to that of erosion and transport of materials (*inert drift*). It involves all benthic organisms, bacteria, algae, and fauna. The size of this drift is linked to the speed of the current.

Part of the drift is of exogenous origin—terrestrial invertebrates that have accidentally fallen into the water. The aquatic drift affects all the organisms exposed to the current, but in different ways according to the species and their developmental stage.

2.1. Forms of drift

There is a constant drift of bacteria, algae, and benthic invertebrates, fortuitous for each individual, but regular along the length of a water course. It is relatively small. Most species drift at particular hours depending on their rhythm of activity (Fig. 14).

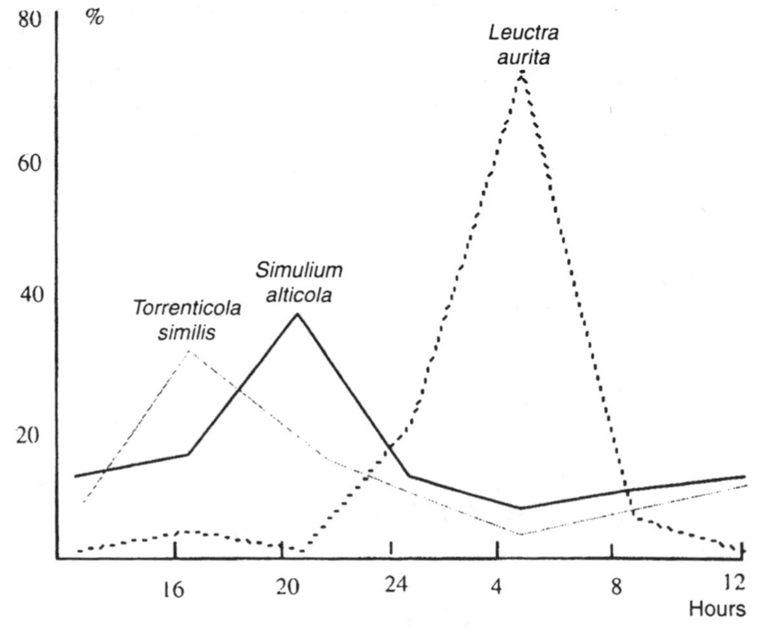

Fig. 14. Rate of drift of benthic invertebrates during the course of a day (Gazagnes, 1983)

Nocturnal drift is the most significant, especially at the beginning or end of the night. It involves species that are inactive during the day and live sheltered from the current. During the night, they move on the surface of the substrate, looking for food (examples are carnivores and creatures that scrape the substrate), and this behaviour exposes them to being carried away by the current.

Small species or those at juvenile stages, which can embed themselves in the alluvia, drift very little, as do species with heavy shells (Trichoptera) and those that adhere to the substrate.

On the other hand, the swimming larvae of Ephemeroptera, Plecoptera (with only a few exceptions), and Crustacea Gammaridae constitute a significant part of the drift. The drift therefore does not exactly reflect the structure of the benthic community (Fig. 15).

In the life of benthic insects with flying adults, emergence is a critical period. There is a risk of being carried away when the adult reaches the water surface, and also during the egg-laying stage. The migrations observed in larvae of Trichoptera *Hydropsyche*, for example, can be more clearly understood in the light of this effect of drift.

Young larvae of the first two stages are abundant near the banks, where the eggs were laid. Larvae of stages 3 and 4 are distributed homogeneously, while older larvae are dominant in deep water, in the middle of the bed. Plecoptera Leuctridae, adapted to life in the interstices of the alluvia by elongated bodies with a reduced cross-section, colonize the sub-flow in the larval stage and practically drift only during their emergence into adulthood.

In a rapid current, older larvae drift more than the young ones, which can sink into the sub-flow. When the current diminishes, the older larvae drift less. Conversely, the young larvae of insects, the Crustaceae Copepoda and Ostracoda, or the Oligochaeta, rise from the sub-flow towards the surface and thus form a significant part of the drift. The abundance of insect larvae in the sub-flow fluctuates during the course of a year. In the stream at Estaragne, in the central Pyrenees, larval density in the sub-flow is highest at snow melt, is reduced during high water, and then increases again with the appearance of a new generation.

Subsequent to experiments on an English stream, J.M. Elliott distinguished three groups of invertebrates according to their mode of drift:

- Some, such as Turbellaria *Polycelis felina*, Mollusca *Ancylus fluviatilis*, Coleoptera Elmidae or Diptera Chironomidae, are transported as inert drift and return to the bottom only by chance.
- Others, such as Ephemeroptera *Rhitrogena semicolorata*, Plecoptera *Protonemura meyeri*, *Lectra*, and *Chloroperla*, and Diptera Simuliidae, are transported as inert drift for currents faster than 19 cm/s. At low current speeds (10–12 cm/s), the percentage of return to the bottom is high and the distance of drift is reduced. On the Neste d'Aure (in

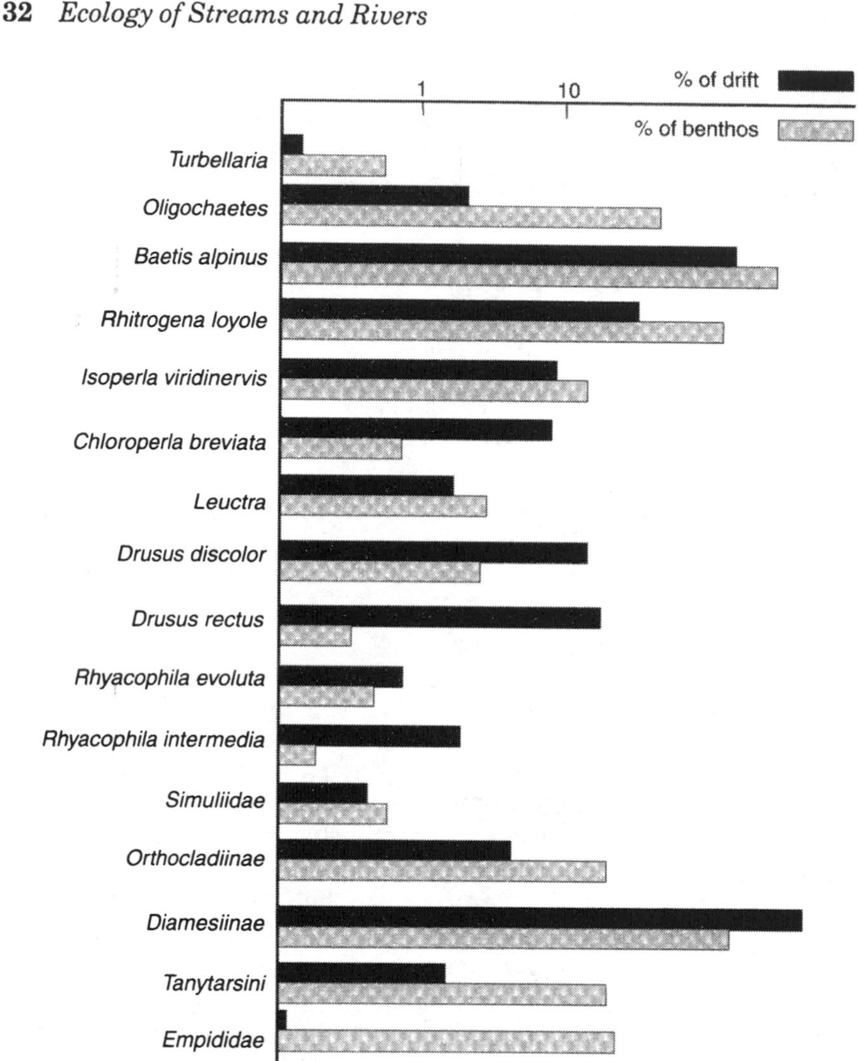

Fig. 15. Annual structure of benthic and drifting populations on the Estaragne stream, upper Pyrenees (Lavandier, 1979)

the upper Pyrenees), at a rapid current various groups of Simuliidae are found in the drift at more than 10 km downstream of their distribution limit.

— A third group is made up of species with a short drift distance and a very high percentage of return to the bottom. This is a drift by successive leaps, in Hirudinea *Erpobdella octoculata*, Ephemeroptera *Ecdyonurus venosus*, *Ephemerella ignita, Baetis rhodani*, and Trichoptera *Hydropsyche*. This mode of drift seems habitual in

Amphipodes Gammaridae: drift of 2.5 m/d in *Gammarus fossarum*, 5 to 50 m/d in *G. zaddaki*.

2.2. Floods and drift

In a period of low water, it is the behaviour of species and their rhythm of activity that essentially determine whether and how they are carried away. In a period of high water, the violence of the current plays a preponderant role, causing instability of the substrate. High water periods may be regular (winter high water, snow melt) or aperiodic (summer floods, release of water from reservoirs).

A flash flood has what can be called a surprise effect on the benthic fauna, notably through the shock wave, which spreads more rapidly than the flood itself during the sluicing of the reservoir. This is manifested in a temporary increase in the drift, which subsequently slows down again. High waters correspond to a phase of reduced activity for the organisms: those in the juvenile stages and small species dig themselves into the sub-flow, the others find shelter in areas protected from the current.

For a pebble of a given diameter, there is a critical speed of current that lifts and moves it, in suspension or by rolling (see Chapter 1). It is thus possible to find out, as a function of the width of the bed and the slope, the flow that can destabilize the substrate and move the fauna (catastrophic drift). On the Miribel canal (upper Rhone in France), which is 80 m wide, a flood of 300 m^3/s carries a catastrophic drift: three-quarters of the invertebrates were transported during the October 1982 floods. Following a severe flood on the Danube, the benthic biomass fell from 5 g/m^2 on average to 0.04 g/m^2. On a stream in the upper Pyrenees, an exceptional flood in 1974 (flow double that of normal high water resulting from snow melt) led to the disappearance of 37% of the trout stock.

2.3. Assessment of drift

A river can be considered a rolling carpet that constantly transports materials and organisms from upstream to downstream. Organisms must strive to avoid being swept away and to compensate for the movement of the current.

The average density of the drift is about 1 individual per cubic metre. However, on the Neste d'Aure, it reaches 7.6 individuals per cubic metre in July-August for Oligochaeta, Hydracarids, Ephemeroptera, and Plecoptera. On the French upper Rhone, the density of the drift is 0.72 individuals per cubic metre on a yearly average—more than 1.2 in summer. If the depth of the Rhone is taken into account, there are 7 individuals/m^2 that are constantly in the water column above the bottom. This represents a drift intensity of more than 58 million invertebrates passing each day in a saturated section, during the summer. The rate of daily renewal of benthos in place is about 0.1 to 1%, depending on the flow, and the drift consequently plays a fundamental role in the functioning of running-water ecosystems.

In the Estaragne stream, at between 2100 and 2370 m altitude, the ratio between the drift and the rate of production could be calculated (Fig. 16). The significance of this drift indirectly indicates the need for mechanisms to compensate the sweep of the current—active upstream movements of the fauna.

These active movements in the substrate have been studied particularly in Amphipodes Gammaridae. The number of individuals that move against the current in 24 h is twice as high as the number of individuals that drift. However, in the aquatic larvae of insects, movement upstream does not go beyond 5 to 6% of transport by drift. It is the movement by flying adults that compensates for larval drift.

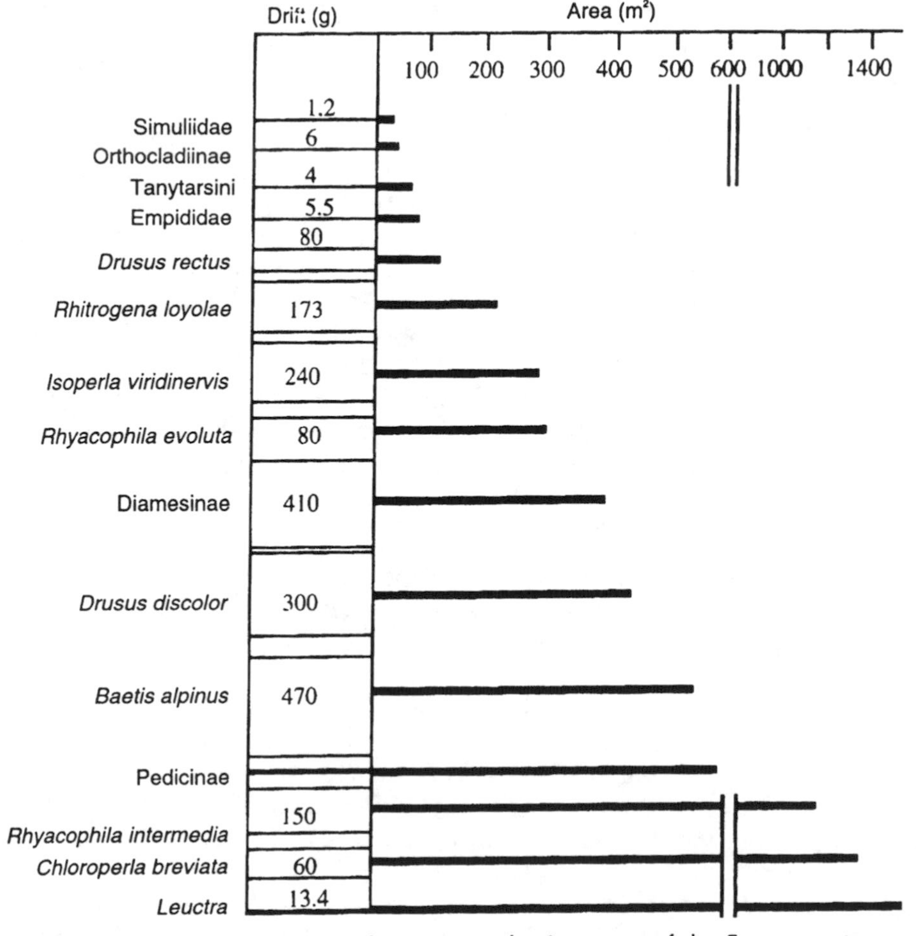

Fig. 16. Significance of drift in relation to production: area of the Estaragne stream required to ensure a production equivalent to the annual drift (Lavandier, 1979)

In 1954, Müller proposed the hypothesis of a cycle during which flying adult females move upstream to lay eggs, thus compensating for the downstream transport. This hypothesis was verified, notably in Ephemeroptera, Plecoptera, and Trichoptera. On the Estaragne stream, the proportion of adult females was higher upstream than downstream, while the sex ratio at emergence of adults was close to 1. On another stream in the central Pyrenees, at an altitude of 2750 m, in very cold waters, only female adults and larvae of *Baetis alpinus* and *Rhitrogena loyolae* were seen, and no older larvae were reported. The larval drift prevents the population of these two species from maintaining itself till the completion of the cycle (too long at a temperature of 3°C). The site is reseeded each year by migrating females that return up the valley. In the Estaragne stream, the females of *Baetis alpinus* and *Isoperla viridinervis* (Plecoptera) are smaller on the lower course (1900 m), whereas all sizes are represented above 2150 m.

One consequence of the drift, on a water course, is a constant tendency to homogenization of benthos and recolonization of the environment. The effect of homogenization is particularly perceptible during high water.

On the Neste d'Aure, for example, at the end of high water resulting from snow melt, the distribution of benthos on the longitudinal profile is relatively homogeneous. It is regulated during the summer. High-altitude species disappear from the lower part of the stream. Ephemeroptera Baetidae progressively abandon the reaches with a slow current, in which Oligochaeta, Crustacea Copepoda and Ostracoda, and Diptera Chironomidae become dominant. On the upper Rhone, the fauna of marginal annexes of the stream (Isopoda *Asellus aquaticus*, some Trichoptera) is washed away by a minor but sudden flood and redistributed in the navigable section.

Recolonization by the drift has been studied experimentally. On the Estaragne stream, a section 5 m long was isolated from the bottom by a plastic sheet covering pebbles and gravels that had been previously washed. The colonization of the new substrate was the result of constant immigration and emigration. Immigration was dominant in the beginning (40 to 80% of the drift is held up in the section, depending on the species). Emigration also began with the colonization and rapidly increased, reaching on an average 75% of individuals from the fourth day (Fig. 17).

Drift therefore seems to be a factor of erosion, transport, and redistribution of organisms as well as inert materials. It is consequently a factor of constant recolonization. When superficial flow is re-established in Mediterranean streams that dry up in summer, the drift from the upper course and points that remain submerged contributes to the establishment of primary communities. This important role of reseeding, in very high-altitude streams as well as in temporary waters, is characteristic of pioneer ecosystems.

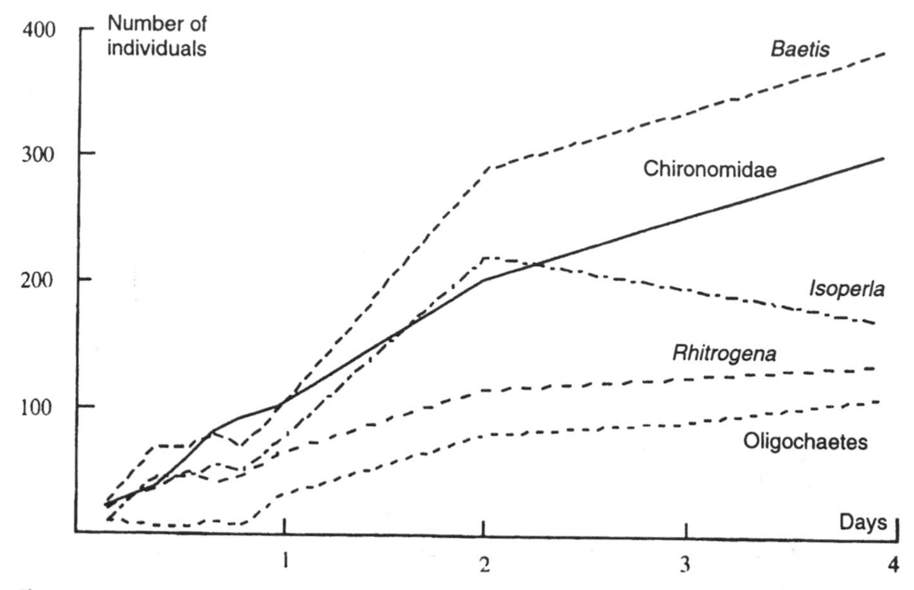

Fig. 17. Processes of colonization of an artificial substrate in the Estaragne stream, upper Pyrenees (Lavandier, 1979)

3. Origin of superficial benthic population

At the water-substrate interface, the image we have of benthos corresponds at each point to an instant photograph of organisms moving from upstream to downstream, from the algae that constitute a biological cover of the substrate up to the invertebrates.

At each point, the survival of a population results from an equilibrium between rates of immigration and emigration, and between the rates of multiplication and loss by drift. It is this latter equilibrium that was expressed by Margalef (1990) in the following form:

$$dN/dt = rN - vp^{(dN/dx)}$$

where N represents the quantitative mean of the population density, r its rate of increase, p the probability (between 0 and 1) for each individual of being carried away by the current, v the average rate of the current (vp being the inevitable drift of the population), and dN/dx the density gradient of the population according to the longitudinal profile of the stream.

To maintain a population despite the losses due to drift, several strategies are possible. The value of p is diminished by the following strategies:

— fixation on the substrate (Diptera Blephariceridae and Simuliidae);
— maintenance in a place sheltered from the current and counter-current movements;
— migration towards the banks and sub-flow.

The value of p is compensated by the following strategies:

— high rate of reproduction (by number of eggs laid or, as in Oligochaeta Naididae, for example, rapid asexual reproduction by individuals);

— return upstream of adult females by air (or even, in the Hydracarids, temporary fixation of larvae on flying adult insects).

The study of benthic organisms of the superficial horizon of the substrate can be envisaged in two different ways:

— On a small scale, the location of each species on the substrate. Here there is the image of a mosaic of microhabitats, each having its specificity and its community.

— On a large scale, the entire course of a river. Here there is the image of a continuum of a river continually eroding its substrate and its benthos. The problems posed therefore concern compensatory adaptations, horizontal migrations and migrations on the sub-flow, continual emigration and immigration, rate of reproduction, and recolonization by airborne dispersal.

In this chronically unstable environment that is the superficial horizon of the substrate, the dominant invertebrates are larvae of insects that become flying adults, capable of continually recolonizing lost habitats by dispersal outside the water. The major role of reseeding is one characteristic of colonizers of pioneering ecosystems.

When the current slows, the magnitude of the drift is reduced. When an aquatic vegetation develops (mosses and macrophytes), it offers shelter from the current. The result is the appearance of invertebrates such as Oligochaeta other than the Naididae, Mollusca, Crustacea Copepoda, and Ostracoda, as well as insects that become aquatic adults. These are forms that would find it difficult to survive in the superficial horizon of the substrate subjected to significant drift, since they lack the potential for recolonization by flying adults. From upstream to downstream, there are two types of aquatic communities that succeed: that of the rhithral characterized by significant recolonization by flying adults, and that of the potamal, where forms lacking the flying stage are highly developed.

In the sub-flow, survival is ensured, except in cases of catastrophic floods, which are relatively unlikely. Fauna that can circulate between the interstices benefits from a stability of the environment on a long time scale. This explains the presence of species with a low rate of growth and lacking a flying stage of dispersal and recolonization.

5

The Hyporheic Environment: Continuity of the Substrate

The hyporheic environment is made up of the interstices of alluvia deposited on the river bed. It is a more stable environment than the superficial horizon of the substrate: the deep layers are reshaped only during exceptional floods. It constitutes an interface or *ecotone*—a zone of contact between the surface and ground waters, between benthic organisms and those of the water table (Fig. 18). The diversified and heterogeneous population reflects this double origin.

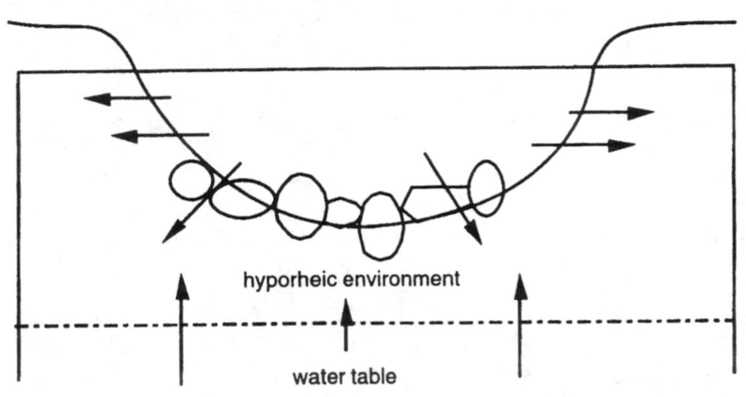

Fig. 18. The hyporheic environment and its water inflow from the river and from ground waters

The surface water penetrates the hyporheic environment by infiltration and, at the thresholds, by a sub-flow. It carries organic materials, which are a source of nutrients, especially during high water. Thus, a biological layer is formed that is inhabited by organisms ranging from bacteria to invertebrates.

1. Hyporheic fauna of superficial origin

Hyporheic fauna is characteristic of this interstitial environment but originates from the superficial horizon—a deep benthos but without particular adaptations. This fauna finds a continuity in this environment that is lacking in the superficial horizon. The deep benthos is the equivalent of endogeal fauna in the terrestrial environment.

The basic population testifies to the continuity of the substrate: Nematoda, Oligochaeta, Crustacea. These are invertebrates with an entirely aquatic cycle, and consequently they cannot recolonize by the aerial route to compensate for drift. Their small size or elongated and supple form facilitates movement between the gravel interstices. This hyporheic fauna is a mixture of species, some strictly or preferentially interstitial (stygobial or stygophilous) and others epigeal (stygoxenous).

To this permanent population is added a temporary population of invertebrates that have an airborne phase in the adult stage (insects) or larval stage (Hydracarids). The juvenile stages of Ephemeroptera, Plecoptera, Trichoptera, larvae of Diptera, and nymphs and adults of Hydracarids are dominant. As a rule, they can be considered stygoxenous because they do not complete their entire biological cycle in the hyporheic environment. However, Ephemeroptera of the genus *Thraulus*, Plecoptera *Leuctra*, with a spindly and supple body, or Hydracarids *Torrenticola*, which penetrate deep into this environment to which they are strictly constrained at certain stages of their cycle, are in fact true stygophiles. A number of species of Hydracarids, shorter at 0.5 mm and lacking pigmentation, colonize only the hyporheic environment.

2. Stygobious fauna of subterranean origin

Stygobious fauna of subterranean origin consists of species that can colonize the hyporheic environment and the deep ground water, and that have characteristics of life in darkness (depigmentation, anophthalmy). Three evolutionary trends are found:

- One is the reduction of metabolism. The oxygen consumption of Amphipoda *Niphargus*, for example, is one tenth that of benthic Amphipoda Gammaridae.
- The second is the reduction of the number of eggs, in relation to an increase in the quantity of vitellus and an elongation of the biological cycle: two to three eggs in *Iberobathynella* (Crustacea Syncarid), two on average in *Microcharon*, and nearly always a single egg in *Angeliera* (Isopoda Microparasellidae). This tendency indicates a perennial environment, without the catastrophic causes of mortality that the benthos face. Natural selection thus favours species with a lower rate of reproduction but a longer life span.
- The third trend is the reduction or disappearance of seasonal rhythms of reproduction, in relation with the low seasonal amplitude of temperature in ground water.

Certain stygobious organisms are characterized by their thigmotactism, i.e., a behaviour that brings them into constant and the widest possible contact with the substrate. If the individual is in contact with a single grain of sand, it cannot successfully avoid it. If the grains are numerous, the generalized contact facilitates the coordination of movements and the individual moves in all directions. Without contact with grains, the individual dies—apparently of exhaustion.

Archiannelids such as *Troglochaetus*, Crustacea Syncaridae (*Iberobathynella*), and Isopoda (Microparasellidae) are thus closely linked to the interstitial environment, to grains of the substrate, and have a supple, spindle-shaped body (Fig. 19).

Other stygobious species have a wider ecological tolerance. Amphipoda *Niphargus* colonize all environments into which light does not penetrate: large lake beds in the hyporheic environment, ground waters, and caves.

3. Distribution of fauna in the hyporheic environment

Two essential factors determine the distribution of fauna in the hyporheic environment: the origin of waters (surface or ground) and granulometry of the substrate.

3.1. Origin of waters

The origin of waters, surface or ground, as well as their possibilities of exchange determine the major lines of the hyporheic population.

In a specially prepared channel of the Miribel canal, upstream of Lyon, a bank of gravels receives the waters infiltrating from the Rhone, carrying with them particulate organic matter. Downstream, it is the ground waters of the alluvial plain that penetrate the hyporheic environment.

Only a small number of species colonize the entire bank, over a little more than 1 km—including Plecoptera *Leuctra* and Amphipoda *Niphargus rhenorhodaniensis* (stygobious).

The area infiltrated by Rhone waters is characterized by a set of benthic species—Oligochaeta, Turbellaria, Microcrustaceae, and Amphipoda Gammaridae, larvae of insects (mostly Chironomidae). *Niphargus casparyi* is the only stygobious species with *N. rhenorhodaniensis*. Both have a wide ecological tolerance.

Benthic species become rare, then disappear in the zone of ground water infiltration, while some stygobious species appear, including *Rhizodriloides* (Oligochaeta), *Parastenocaris* (Copepoda), *Pseudocandona* (Ostracoda), *Niphargus kochianus* and *Salentinella* (Amphipoda), and *Proasellus* (Isopoda).

The confluence of the Guiers with the Rhone is accompanied by an increase in infiltration of ground water in the hyporheic environment, where it mingles with waters of surface origin. The fauna therefore behaves simultaneously like benthic and stygobious species. However, in the dead

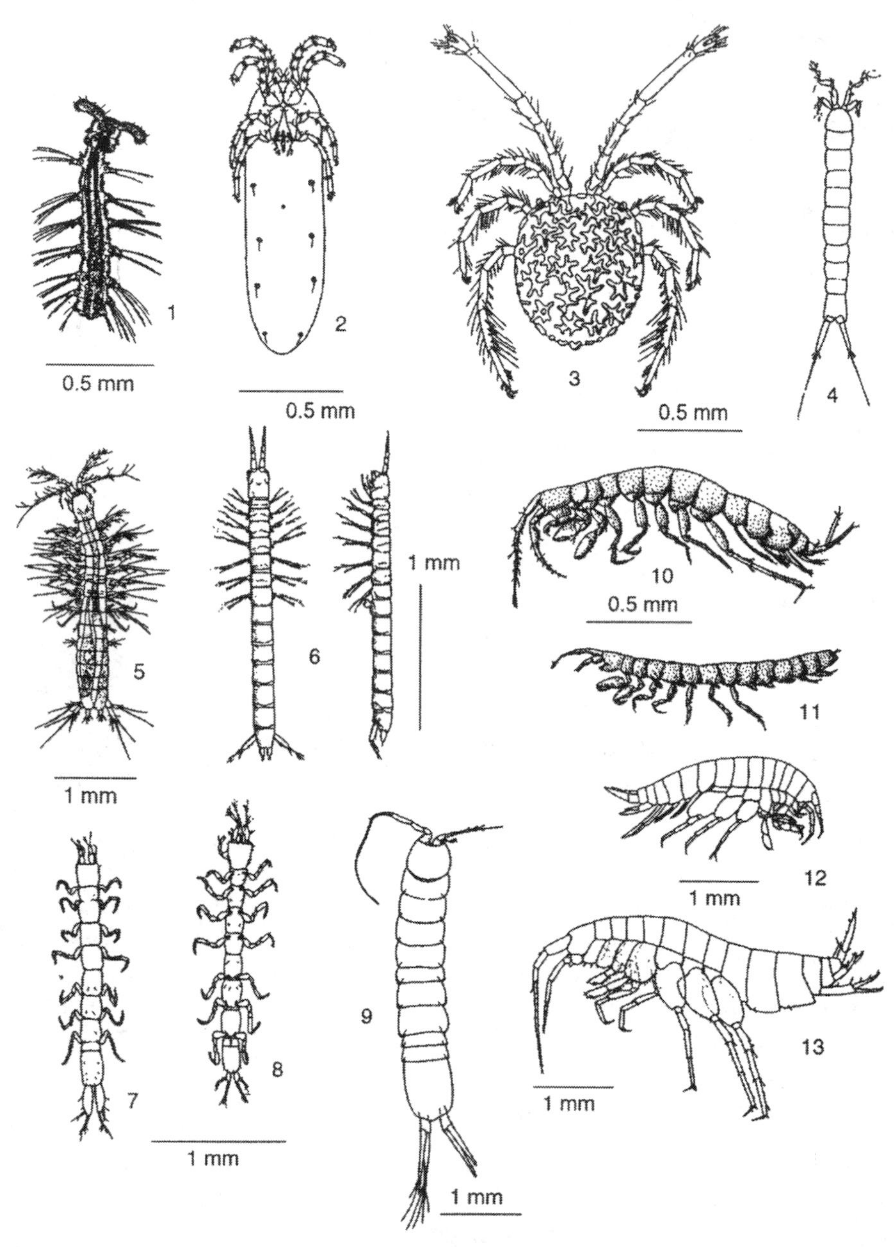

Fig. 19. Characteristic invertebrates of a hyporheic environment. 1. *Troglochaetus bernecki*, Archiannelid. 2. *Parawandesia chappuisi*, Hydrachnellae. 3. *Momonides lundbladi*, Hydrachnellae. 4. *Parastenocaris*, Copepod Harpacticid. 5. *Bathynella natans*, Syncarid. 6. *Iberobathynella fagei*, Syncarid. 7. *Microcharon*, Isopoda. 8. *Angeliera*, Isopoda. 9. *Stenasellus*, Isopoda. 10. *Bogidiella albertimagni*, Amphipoda. 11. *Ingolfiella acherontis*, Amphipoda. 12. *Salentinella angelieri*, Amphipoda. 13. *Niphargus jugoslavius*, Amphipoda (Angelier, 1959; Henry and Magniez, 1963; St. Karaman, 1954; Ruffo, 1951, 1959).

branches of the Rhone, in the hyporheic environment supplied solely by ground water, the population is essentially of stygobious origin.

It is thus the origin of water that determines the structure of the population in the hyporheic environment. However, this origin may vary throughout the year. The hyporheic zone is recharged from the surface water when the water table drops—and from ground water when the table rises again. The population structure varies as a function of the origin of the water, by vertical migrations of the benthic and stygobious fauna.

During severe dry spells, infiltration of surface water is insufficient; the oxygen concentrations are low and the environment more reducing. The stygobious species are relatively less sensitive to low oxygen concentrations but the more sensitive benthic species move towards the superficial horizon. In the streams of the northern Alps, for example, Hydracarids of the genus *Torrenticola* essentially colonize the hyporheic environment, in well-oxygenated waters. Along the Mediterranean coast, the same species are limited to the surface horizon during the summer.

In temporary streams, certain benthic species find refuge during the dry period in the hyporheic environment, in the active stage of life (Turbellaria, Isopoda Asellidae, Coleoptera Elmidae) or less active stage (Mollusca *Ancylus*).

Apart from vertical migration, hyporheic fauna also migrate horizontally. Horizontal migrations result from seasonal variations in the surface water level, with immersion or emersion of the substrate.

3.2. Granulometry of the substrate

There is a close correlation between the distribution of hyporheic fauna, its diversity, and the granulometry of the substrate. The granulo-metry determines porosity, *i.e*, the percentage of spaces occupied by water in relation to the total volume of sediments.

For a given species, it can be said that there are two types of sediments: those in which the particle diameter determines the interstices allowing circulation of individuals and finer sediments that seal them off. The maximum faunal diversity corresponds to high proportions of grains of 2 to 12 mm diameter and less than 5% of grains of diameter less than 0.2 mm.

The percentage of grains of less than 0.2 mm has a decisive importance in the distribution of interstitial organisms. An excess of sand, clay, and silt makes the lacunary spaces uninhabitable for interstitial species that normally move about in the spaces. These species are replaced by burrowing species, which progress by moving the grains. The richest interstitial fauna is found in the middle course of streams (towards the base of the rhithral). In the lower course, the potamal, the burrowing species are dominant when there is a high proportion of clays and silts.

Granulometric requirements can be more precisely calculated. For example, Amphipoda *Bogidiella* and Isopoda *Angeliera* colonize coarse, non-homogeneous alluvia with a maximum of grains of 0.8 to 1.6 mm. In the sub-flows of the upper Moroccan Atlas region, Amphipoda *Metacrangonyx knidiri* is less tolerant of a high proportion of clays. *Metacrangonyx spinicaudatus* lives in a substrate with a proportion of clays lower than 20% and with a proportion of gravel lower than 56%, unlike *M. knidiri* (Fig. 20). Isopoda *Microcerberus* colonizes a vast and heterogeneous granulometric spectrum, with less than 20% clay—the average values of grains range from 0.57 to 2.5 mm, and the gravels in proportion lower than 52%. *Microcharon*, another Isopoda, prefers coarse sand of 0.9 to 1.15 mm, with less than 20% of clays and 36% of gravels.

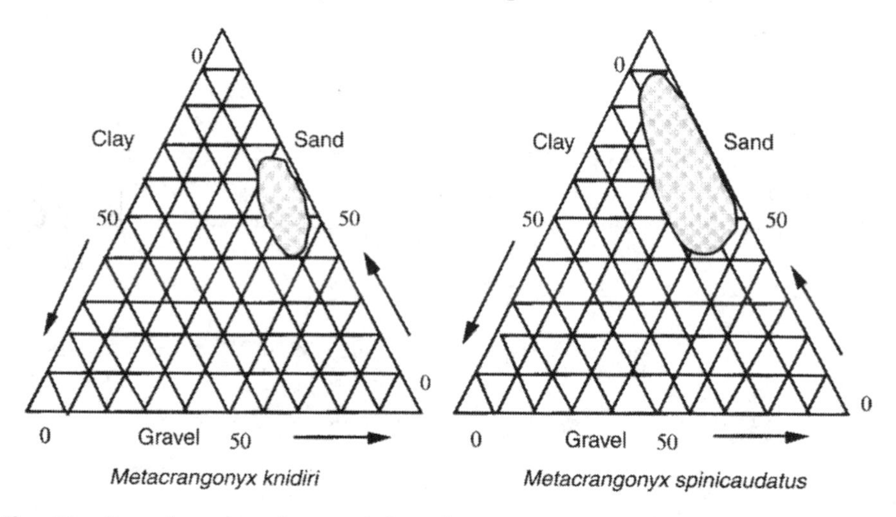

Fig. 20. Granulometric characteristics of two environments with *Metacrangonyx* (Amphipoda) of Morocco (De Bovee et al., 1995)

The granulometry of sediments reflects the conditions of flow of surface water and the speed of the current. On a single sector of a stream, the current determines zones of undermining and deposits with variable characteristics of granulometry, porosity, and levels of oxygen in the water. The distribution of organisms in the sub-flow is a function of these characteristics.

On the Lachein stream (in Ariège, France), the distribution of about 50 species of Crustaceae, of which 22 are stygobious, was studied on a section of 75 m^2 (Fig. 21). The sub-flow was supplied by surface as well as ground waters.

The surface flows determine four zones: two gravel banks, a channel, and a dead water zone. The upstream part of the gravel banks is characterized by a high percentage of sand and grit of 1 to 8 mm diameter, and some gravel. In the channel, there are gravel and pebbles of 25 to 27 mm diameter. Downstream of the gravel banks the granulometry is intermediate between that of the upstream part of the gravel banks and

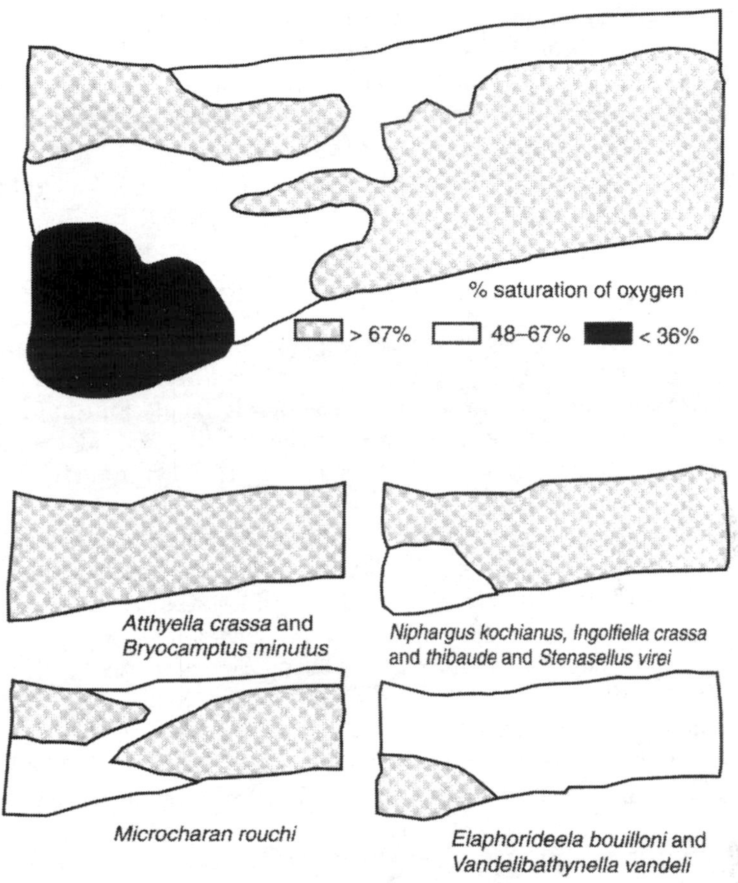

Fig. 21. Distribution of some Crustaceae in the hyporheic environment of the Lachein stream (Ariège) as a function of the substrate and dissolved oxygen concentration (Rouch, 1988)

that of the channel. The highest porosity is found in the upstream part of the banks. The lowest concentrations of dissolved oxygen (less than 36% of saturation) are located in the interstitial part of the dead water area.

A Copepoda, *Elaphoidella bouilloni*, and a Syncarid, *Vandelibathynella vandeli*, both stygobious species, have a limited distribution in the dead water area. Two Copepoda, *Ceuthonectes gallicus* and *Parapseudoleptomesochra subterraneus*, as well as Isopoda *Microcharon rouchi*, extend over the gravel banks, while Copepoda *Bryocamptus minutus* (a benthic species) occupies only the periphery.

Two benthic Copepoda—*Bryocamptus echinatus* and *Attheyella crassa*—are established in the channel zone. Amphipoda *Salentinella* and *Parasalentinella* also occupy this zone, but in addition they extend to the periphery of the gravel banks.

Analogous observations were recorded on the Ostracoda in Austria and in the alluvial plain of the Ain (France).

The origin of waters of the hyporheic environment, surface or ground, determines the major groups in the hyporheic population. On a smaller scale, however, the surface current gives rise to a mosaic of microhabitats. The result is a microdistribution of organisms, similar to that in the superficial horizon of the substrate (see Chapter 4).

4. Origin and biogeography of stygobious fauna

In the surface waters, problems of biogeography often appear secondary in the comprehension of the distribution of organisms. The upstream-downstream continuum of streams and the possibilities of faunal dispersal (flight, swimming, drift) ensure a fairly wide distribution, at least on the scale of a watershed. Ecological factors are of primary importance in the distribution of organisms.

It is not the same for stygobious fauna. A valley such as the Rhone valley, for example, is a series of alluvial plains separated by passes in which the hyporheic environment is reduced, if not absent. Ground waters seem like discontinuous spaces isolating species with poor dispersal capacity. The distribution of stygobious fauna cannot be understood from just the prevailing ecological factors. Historical factors must also be considered—origin of the species, conditions of colonization of the subterranean domain, and isolation of populations. The present stygobious fauna has a double origin: surface waters and the marine interstitial environment.

4.1. Stygobious fauna originating from surface waters

The major stygobious fauna originating from continental waters are Turbellaria Triclades, Mollusca, Crustacea Cladocera, Cyclopids, and Hydracarids (Hydrachnellae). Oligochaeta, Crustacea Ostracoda, and Harpacticids, while being mostly of continental origin, include some species of marine origin.

Crustacea Syncarids are known as fossils of marine, brackish, or fresh waters of the Carboniferous and Permien eras. Only species of continental waters exist today, nearly all of them stygobious. They are very old relics. Some believe them to originate from surface waters; others consider them to be marine stygobious fauna that has become continental by successive waves starting from the Secondary era.

Around the Mediterranean basin, outside the Quaternary periglacial area, an old lineage of Isopoda Asellidae—of the *Proasellus meridianus* group—presents all the intermediates between pigmented and eyed species and depigmented and eyeless species, with a high rate of endemism. However, species of the northern Alpine and periglacial area have a wider distribution: from Wales to Provence in France for *Proasellus cavaticus*, which probably recently colonized the subterranean field (post Ice Age). *Asellus aquaticus* is a species of surface waters. It has stygobious populations and subspecies, partly depigmented and eyeless. The same holds true for certain Amphipoda Gammaridae.

Among the Hydracarids, 11 sub-families no longer have a representative in surface waters and are considered very old stygobious fauna. However, in many genera (*Torrenticola, Atractides, Feltria, Aturus*), benthic or muscicolous species and other hyporheic fauna are found. The genus *Torrenticola* is characteristic of the surface waters of southern Europe, but it is stygophilous in the northern Alps. The same is true for *Bandakia corsica* and *Axonopsis vietsi*. This indicates a recent or current colonization.

The stygobious fauna of surface continental waters thus presents all the modes of colonization and types of adaptation since the ancient stygobious lineages, unrelated to present benthic fauna (Syncarids), and recent colonizers with few real adaptations.

4.2. Stygobia of marine origin

Apart from some species of Oligochaeta, Ostracoda, and Copepoda, most stygobious Isopods and Amphipods have marine ancestors that first colonized the coastal interstitial environment. During marine regressions, certain populations followed the retreat of the marine coast. Others remained in place, progressively adapting themselves to continental waters. Since the duration of a marine regression corresponds to thousands of generations, the change of environment was insignificant for a single generation.

The marine interstitial environment is less stable than the deep continental subterranean environment, and species transformation in the former is more rapid. The very old continental species therefore have more primitive characters than recent species and their counterparts that are still marine.

In France, the distribution of Isopoda Microparasellidae of the genus *Microcharon* corresponds to marine regressions of the Tertiary era. The most primitive species colonize the regions reached during the Cretaceous period and during the Eocene period by the Tethys—the gigantic depression, distant ancestor of the present Mediterranean Sea—which subsequently remained constantly emerged. The most evolved species colonized the coasts of the Pliocene sea. About ten species of the genus *Microcharon* are marine. They are characterized by their eurythermy and euryhalinity.

Isopoda Microcerberidae, Stenasellidae, and Amphipoda of the genus *Pseudoniphargus* present the same type of distribution and evolution as Microparasellidae.

Amphipoda of the *Niphargus* group are phyletically connected to the marine *Eriopisa* and *Eriopisella*, of which several species are interstitial.

Stygobial fauna of marine origin is thus different from stygobial fauna of continental origin. There are a few stygophiles of intermediate forms, but the distribution of some stygobious fauna reflects marine transgressions and regressions over the course of geological eras.

5. Conclusions

The hyporheic environment seems effectively to be an ecotone, a zone of contact between two worlds, one benthic and the other subterranean. The inputs of organic matter from the surface water constitute the source of nutrients common to the two worlds and are the origin of the biological layer.

Despite these organic inputs, the hyporheic environment does not appear to be a true soil. In a terrestrial soil, there is a total cycling of materials of the ecosystem. The hyporheic environment receives only a small part of these materials, since their flow is horizontal, upstream to downstream, and not vertical as in a soil. The hyporheic environment plays only a minor role in the functioning of a running water ecosystem. A true soil appears only on the lower course, in the potamal in a slow current, where there is a massive deposit of organic and mineral matter.

The greater continuity of the substrate in the hyporheic environment than in the superficial horizon makes it a refuge in which benthic invertebrates are cut off from both the drift and high waters. Movements by this benthic fauna in the depths are in fact observed in high water, followed by a rise to the surface. However, this rise towards the superficial horizon generally is not of great importance in the recolonization of disturbed zones.

The hyporheic environment can be considered an additional environment of the water course, with its own functioning, continuity, and problems of endemism. However, its importance in the stream ecosystem has been underestimated in quantitative terms with respect to the benthos. The distribution of benthos must be studied no longer on the basis of just the surface horizon, but also within the thickness of the sediments; a volume of substrate must bé considered, and not just an area. This is a partial explanation of what is called the *Allen paradox*, according to which the consumption of invertebrates by a population of trout represents 6 to 40 times the benthic stocks. One explanation lies in our limited understanding of the rate of renewal of these stocks and the intensity of the drift. Another lies in the very method by which these stocks are estimated, based on just the surface horizon and not on the surface and interstitial environment taken together. It is in this field that the hyporheic environment plays an important role in the functioning of the stream ecosystem.

6

Macrophytes of Running Waters: A Substrate for Algae and Fauna

Apart from benthic algae and lichens, the plant life of running water is made up of macrophytes: Bryophytes, Pteridophytes, and Spermatophytes. They are characterized by their various modes of anchoring themselves on the substrate, which determine their ability to resist the current. As for benthic fauna, the current determines the population, directly or by way of the substrate, but light also plays a significant role. Macrophytes in turn constitute a substrate for algae and fauna and help to protect them from the current.

1. Bryophytes

Bryophytes of running water are essentially mosses as well as Hepatica. Their stems have *rhizoids* at the base, or basal discs. Rhizoids are comparable to absorbent hairs on roots. They serve to absorb as well as fix. They are fine enough (about 10 μm diameter) to insinuate themselves into the smallest rock crevice. Their ability to selectively increase their volume enhances their fixation efficiency (Fig. 22).

1.1. Colonization of the substrate

Colonization of a rocky substrate by Bryophytes begins in the water or close to water, in shaded zones that are kept moist by trickles of water or the spray from waterfalls (hygropetric surface), or even under an overhanging rock, sheltered from the current. From the point of implantation, they extend along the substrate, both amphibious and submerged, until the power of the current limits their extension.

Bryophytes grow slowly. At 1500 m altitude, in the Pyrenees, it has been demonstrated that half the plants of *Fontinalis squamosa* are aged 3 to 5 years, and some reach 9 years of age. This underlines the fact that Bryophytes need a stable substrate, a mother rock, and large stones resistant to high water in order to develop. In theory, Bryophytes are consequently confined to the upper course of rivers and streams, in the

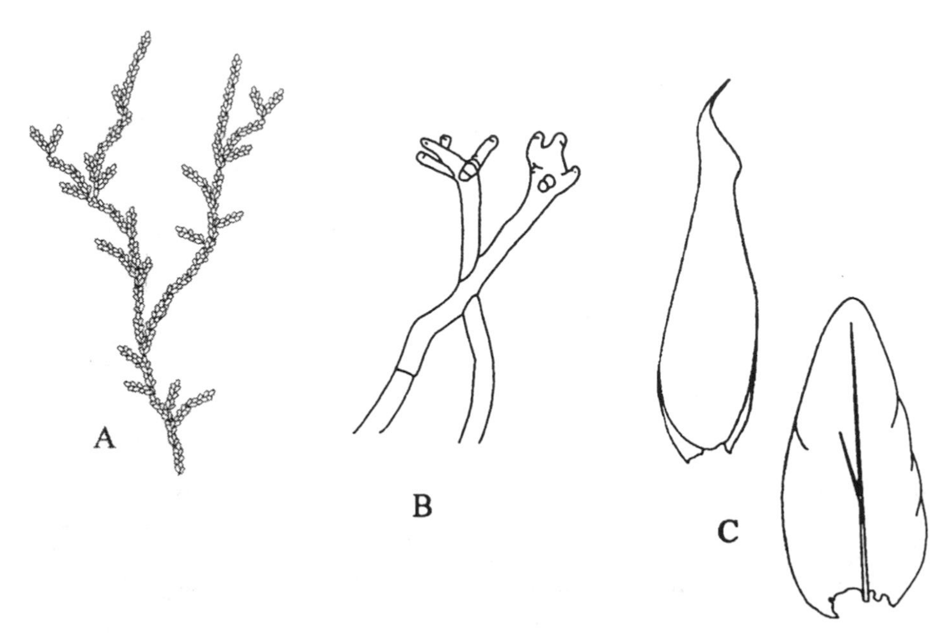

Fig. 22. *Platyhypnidium riparioides* (Bryophyte). A, branch. B, rhizoids allowing anchorage on the substrate (diameter 11 μm). C, boat-shaped leaf, face and profile (Devantery, 1987).

rhithral, which has such substrates. Their presence in the lower course is related to the presence of perennial substrates—stone-reinforced banks, constructed thresholds, bridge piles, stones with diameters greater than the competence of the current, or even roots of trees along the banks.

1.2. Population of Bryophytes

Bryophytes constitute a habitat for algae and torrenticolous invertebrates, but the protection they provide against the current is highly variable.

Mosses of the group *Fissidens*, for example, have short, oblique stems (0.5 to 3 cm on average) with leaves in the form of strips. The stems of *Bryum* are also short and offer moderate protection against the current.

The stems of *Fontinalis*, conversely, are long (10-30 cm), ramified and have wide leaves. In *Platyhypnidium riparioides*, the stems are 4 to 8 cm and longer, also ramified and lying flat. The leaves are 2.5 mm long and have a low height-width ratio; they are boat-shaped and constitute a good protection for fauna against the current.

Fauna is not distributed uniformly throughout a Bryophyte area. The zones above water, drenched by spray, are colonized mostly by larvae of Diptera Psychodidae, Stratiomyidae, Dixidae, Tipulidae, and certain Limoniidae, forms that are more amphibious than aquatic. These are called *madicolous* fauna. Such zones are also egg-laying sites preferred by flying adults of aquatic insects. This explains the presence of very young larvae of Ephemeroptera, Plecoptera, Trichoptera, and Diptera in the Bryophytes. These larvae later migrate towards the submerged strata.

The morphology of Bryophytes and the current determine the population of the submerged zone. When the stems are erect or even leafless, or in a slow current, the invertebrates move more easily. The population is thus made up of Oligochaeta Naididae, swimmers, Hydracarids such as *Panisus, Panisopsis*, Plecoptera, Ephemeroptera, Coleoptera (larvae and adults), Trichoptera such as *Micrasema* and *Drusus*, and Diptera Chironomidae. The fauna is quite similar to those at the base of pebbles and subjected to the same risks of drift, because the protection against the current is limited.

When Bryophytes have long stems, lying flat, with broad leaves, the protection of invertebrates against the current is less certain. Areas of laminar currents may lead to a boundary layer. A real muscicolous fauna thus appears. This includes elongated and supple forms, such as Oligochaeta Enchytraeidae, and especially species of small size, including Crustaceae Ostracoda, Copepoda Cyclopides and Harpacticids, and Hydracarids of the genera *Feltria, Axonopsis, Aturus*, and *Kongsbergia*. They generally do not grow longer than 0.5 mm, and they find protection against the drift between the leaves of Bryophytes as well as in the interstitial environment. Ostracoda and Copepoda are abundant: these purely aquatic organisms cannot recolonize the medium by the aerial route. Bryophytes, which colonize substrates that are stable over time, ensure a perennial habitat for this muscicolous fauna, but they play a selective role in the size of species.

According to R. Margalef's formula (Chapter 4, section 3) concerning the conditions of survival of a population in running water, the average speed of the current is low in the Bryophytes, and the probability of drift is low for muscicolous fauna. Under these conditions, compensation for the drift by airborne recolonization does not seem essential, and the growth rate need not be high. Muscicolous Hydracarids have fewer eggs (1 to 3) than those of the superficial horizon of the substrate, around as many as stygobious species.

2. Spermatophytes with rooted plant life

Rooted plant life may be either entirely submerged, as with certain *Ranunculus* or *Myriophyllum* (Hydrophytes), or partly emerged, such as *Carex, Phragmites*, or *Menyanthes* (Halophytes), which are annual or perennial. These are flexible plants, with generally elongated leaves, which offer minimal resistance to the current. Some of them tolerate a temporary drying out in low water. There is no true hiatus between waterside terrestrial flora and aquatic flora.

2.1. Colonization of the stream environment

Although the Bryophytes are confined to stable, hard substrates, Pteridophytes and Phanerogams are associated with a substrate consisting of small pebbles, gravel, sand, and silt.

The colonization of an environment by vegetation follows a dynamic process resulting from the current and also from the plant-substrate interaction. It leads to the formation of a soil analogous to a terrestrial or lake shore soil.

A stream in the central Pyrenees at 2000 m altitude (the Estibère), for example, runs through a meadow. As the stream enters the meadow, the slope diminishes, and horsetails (*Equisetum limosum*) are the pioneer colonizers of a substrate of gravel and sand, with a current less than 10 cm/s at low water. The roots go deep and the stems of the horsetails lie flat during high water. Horsetails help to slow the current and, close to the banks, *Carex rostrata* can establish itself. The rest of the succession is a function of the depth.

Above 30 cm, there is only *Sparganium angustifolium*, with long floating leaves. Below 30 cm, a Menyanthaceae, *Menyanthes trifoliata*, mingles with *Carex*. Their spread-out and superficial roots thus help to stabilize the substrate.

A high density of *Carex* and *Menyanthes* no longer prevents water from flowing normally across the vegetation. The stream bed contracts, and the current hollows out the bottom to allow the flow to continue. The depth and the current thus fix a limit to plant life expansion.

Algae (periphyton) fix themselves on the stems and leaves of plants. In the vegetation of horsetails and *Sparganium*, where the water circulates, a silt of allochthonous origin, brought in during high water, is deposited during low water. Nitrogen and phosphorous concentrations in the interstitial water are higher than in the surface water, and they are present in oxidized form as nitrates and phosphates. When the water no longer circulates in the dense growth of *Carex* and *Menyanthes*, or within the substrate, organic matter accumulates on the bottom. It is of autochthonous origin, from the decomposition of vegetation in the autumn. Nitrogen and phosphorous concentrations in the interstitial water are very high but indicate an environment that has become reducing (ammoniacal nitrogen). The biosynthesis-biodegradation gradient is vertical, from the surface towards the bottom.

There are consequently three factors involved in the distribution of vascular plants in the Estibère stream: current, nature of the substrate, and depth.

On the Trieux tributary (northern Brittany), the drainage order number reflects an upstream-downstream succession of species (Fig. 23, Table 3). Around the springs, a slow-current flora colonizes the bed (*Montia fontana*,

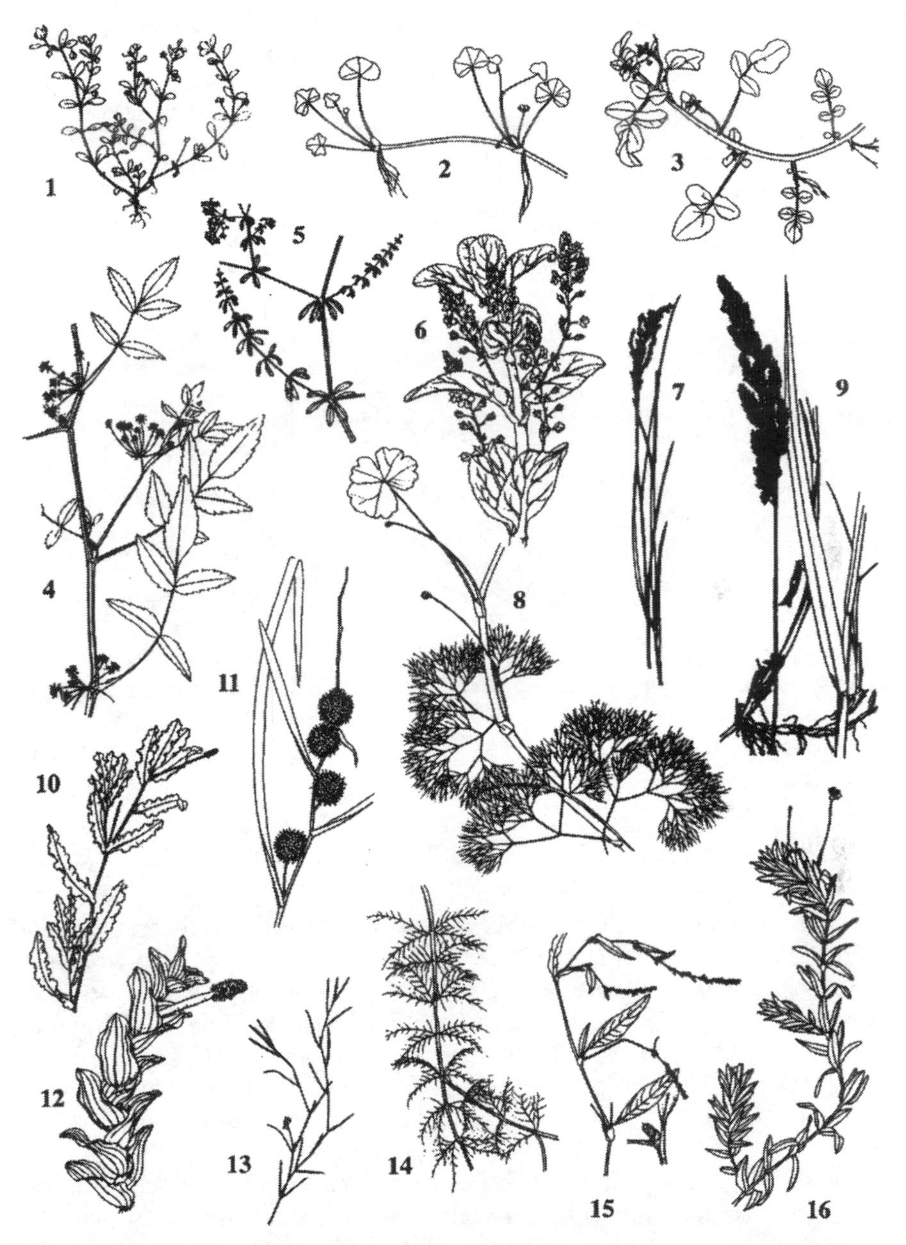

Fig. 23. Phanerogams of the Trieux tributary (Brittany) from upstream to downstream. 1. *Montia fontana*. 2. *Ranunculus hederaceus*. 3. *Nasturtium officinale*. 4. *Apium nodiflorum*. 5. *Veronica beccabunga*. 6. *Glyceria fluitans*. 7. *Galium palustre*. 8. *Ranunculus peltatus*. 9. *Phalaris arundinacea*. 10. *Potamogeton crispus*. 11. *Sparganium emersum*. 12. *Potamogeton perfoliatus*. 13. *Potamogeton pusillus*. 14. *Myriophyllum alterniflorum*. 15. *Polygonum hydropiper*. 16. *Elodea canadensis* (Haury, 1988; Coste, 1937; Hers et al., 1967).

Table 3. Longitudinal distribution of heliophilous phanerogams on the Trieux tributary (Brittany) as a function of order of drainage (Haury, 1988)

	Order of drainage				
	1	2	3	4	5
Montia fontana	+				
Ranunculus hederaceus	+				
Nasturtium officinale	+				
Glyceria fluitans	+	+			
Apium nodiflorum	+	+			
Veronica beccabunga	+	+			
Galium palustre		+	+		
Ranunculus peltatus		+	+		
Scutellaria galericulata		+	+		
Callitriche obtusangulata			+	+	+
Callitriche paniculata			+	+	+
Callitriche platycarpa			+	+	+
Ranunculus penicillatus			+	+	+
Polygonum amphibium			+	+	+
Phalaris arundinacea			+	+	+
Elodea canadensis					+
Sparganium emersum					+
Potamogeton pusillus					+
Potamogeton crispus					+
Potamogeton profaliatus					+
Myriophyllum alterniflorum					+
Nuphar lutea					+
Polygonum hydropiper					+

Ranunculus hederaceus, etc.). This is succeeded by a small-stream flora (orders 1 and 2) consisting of *Glyceria fluitans, Apium nodiflorum*, and other species, followed by a set of rheophilous species (orders 3 and 4, currents of around 0.5 m/s), with notably *Callitriche* and *Ranunculus peltatus*. The bed widens and the vegetation moves towards the banks. On the lower course (order 5), there appear *Potamogeton, Elodea canadensis, Myriophyllum alterniflorum, Sparganium emersum*, and other species.

The substrate of *Callitriche* groups comprises 35 to 55% blocks, stones, gravel, and coarse sand (Fig. 24). However, in the lower course, silt, organic debris, and roots constitute 75 to 90% of the substrates with *Potamogeton* and *Phalaris arundinacea*. By helping to slow the current, macrophytes trap and deposit fine materials. The evolution of the substrate, from *Callitriche* to *Potamogeton*, results from an interaction between the current and plants, the ultimate stage of which is the constitution of a soil, as in the Estibère stream.

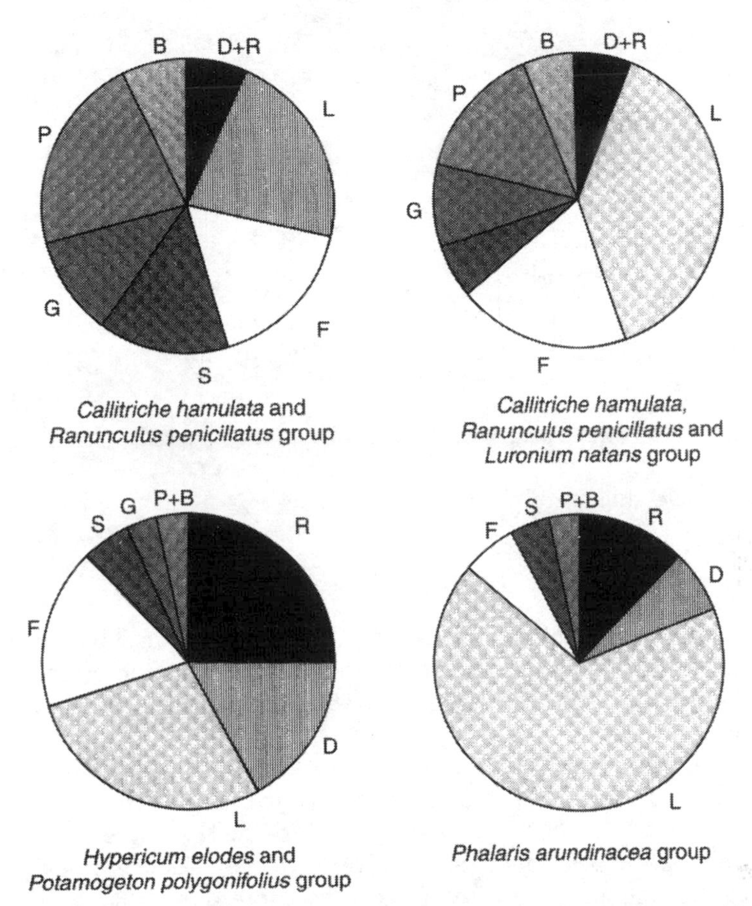

Fig. 24. Nature of substrate of some plant groups in Scorff (Brittany). B, blocks. P, stones and rocks. G, gravel. S, coarse sand. F, fine sand. L, silt. D, plant debris. R, roots (Haury, 1991).

The depth of the bed increases, as a rule, from upstream to downstream. *Glyceria fluitans* colonizes beds of around 20 cm, while *Ranunculus penicillatus* is abundant on bottoms of 40 cm, and *Phalaris arundinacea* and *Nuphar luteum* on those of 60 cm (Fig. 25).

The current structures the population, directly or by the nature of the substrate that it determines. In addition to depth, light and content of dissolved salts are also important ecological factors for plants: light determines their distribution in depth or even their location (shaded or sunny banks), and the dissolved salts select species as a function of the geological nature of the watershed (Chapter 10). However, the ecological and eco-morphological plasticity of certain species must be emphasized, which differentiates aquatic and emerged forms (especially *Ranunculus, Callitriche, Sagittaria*).

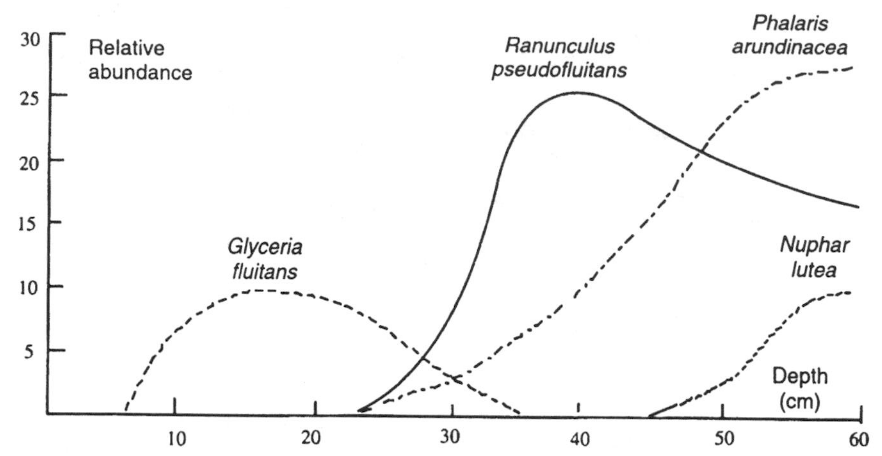

Fig. 25. Relative abundance of some phanerogams as a function of depth (Haury, 1989)

2.2. Fauna of rooted vegetation

Grasses constitute a habitat for the periphyton—algae and bacteria fixed on the vegetation, as well as invertebrates and fish. The interaction between current, vegetation, and substrate generates several types of microhabitats as a function of level, in a single herbarium.

On the Dordogne river, downstream of the Argentat dam, some invertebrates closely linked to the vegetation colonize the *Ranunculus* plant settlement, at all levels, from the base to the extremities of the tufts (the Epheroptera *Ephemerella ignita* or the Cladocera *Eurycercus lamellatus*). Trichoptera Rhyacophilidae and Diptera Simuliidae, which are rheophilous, are located in the median and distal parts of tufts, in contact with the current.

Conversely, the base of tufts and the sediment shelter fauna living in slow or stagnant water, which consists of swimming, benthic, or burrowing organisms: Mollusca *Lymnaea truncatula*, Oligochaeta Lumbriculidae, Enchytraeidae, Tubificidae, and Naididae, Crustacea Ostracoda and Cladocera (*Daphnia* and *Alona*), Isopoda Asellidae, Amphipoda Gammaridae, Plecoptera *Leuctra*, and Diptera Chironomidae.

Some species, which live at the base of tufts in July, when the flow of the Dordogne is greater, migrate towards the median and distal level of the vegetation during low water at the end of the summer: Trichoptera *Brachycentrus* and *Hydropsyche*, Diptera *Paratrichocladius* (Orthocladiinae).

The vegetation also shelters fish. Trout, particularly juveniles, prefer to stay between the tufts, in a slow current. Species such as carp, roach, and pike are more closely connected to the vegetation than Salmonids. There they find a refuge from the current, protection against predatory fish, their food, and a site for reproduction. Carp, bream, bleak, and pike lay their eggs in the vegetation. With the exception of pike, they are omnivores with a carnivorous tendency and find invertebrates they can feed on in the vegetation. But species that are truly herbivorous are rare. Chinese carp, a herbivore of Asian origin, is used in Europe to control the growth of vegetation in closed water bodies.

7

Life in the Water Trail: Plankton

Along with the beds of water courses and the hyporheic environment, the water itself constitutes an environment colonized by organisms living in suspension called *plankton*.

Plankton is a group of small organisms whose own movements are not sufficient to keep them independent of the current. In running waters, therefore, they are constantly swept downstream. Plankton of stagnant waters in themselves practically constitutes an autonomous ecosystem, with primary producers (algae), primary and secondary consumers (rotifers, crustaceans, fish), and bacteria. The water transit time in a lake may be many years. There is therefore an environmental continuity and the ecosystem is organized accordingly. In running waters, the transit time—the time the water takes to move to its confluent or estuary—is shorter, and the continuity of the environment is insufficient for the organization of an ecosystem.

The drift as well as benthic algae and invertebrates attached to the substrate are found suspended in the water (Fig. 26). These organisms survive for a short period and die after some time if they cannot rejoin their substrate, probably by the mechanical action of the current—friction or crossing of thresholds. Similarly, during high water, organisms in dead water branches and floodplains are swept along by the current. This latter phenomenon is particularly notable in water courses with a large floodplain, such as the Amazon or the Orinoco.

True plankton, developing in open water, appears only on the lower course, the potamal, of rivers or on canals. In a river, when the depth is greater than one metre and the current slow, phytoplankton develops over a large area.

1. Transit time and development of plankton

The types of planktonic organisms are fundamentally the same in running and stagnant waters: algae (phytoplankton), protozoa, rotifers, and crustaceans Cladocera and Copepoda (zooplankton). The transit time of the water in the river is the first limiting factor that determines the nature and density of the plankton, as a function of growth rates of each species. Also

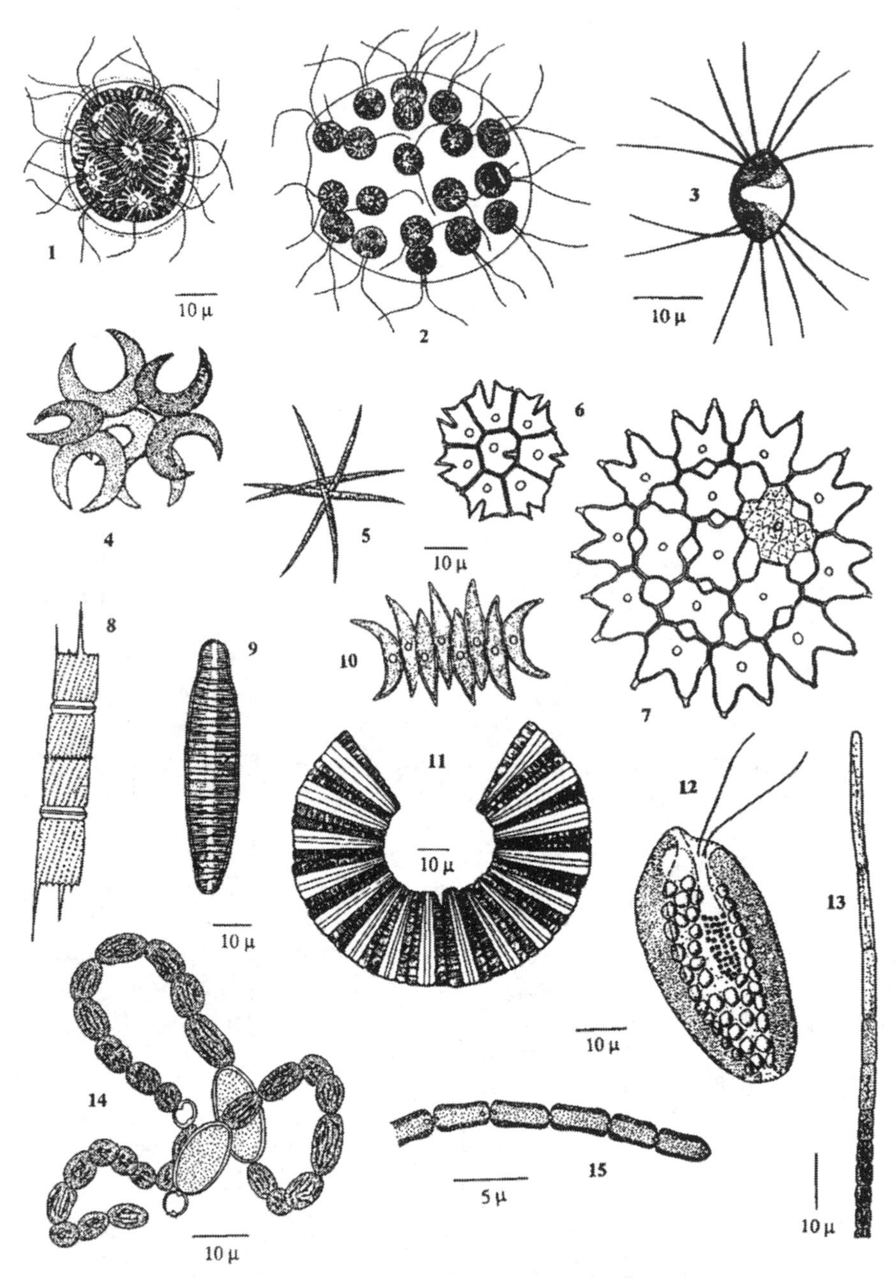

Fig. 26. Types of algae characteristic of running waters—Chlorophyceae: 1. *Pandorina morum*. 2. *Eudorina unicocca*. 3. *Chodatella ciliata*. 4. *Selenastrum bibraianum*. 5. *Ankistrodesmus falcata*. 6. *Pediastrum tetras*. 7. *Pediastrum duplex*. 8. *Scenedesmus falcata*. Diatoms: 9. *Aulacoseria granulata*. 10. *Diatoma vulgare*. 11. *Meridion circulare*. Cryptophyceae: 12. *Cryptomonas ovata*. Cyanophyceae: 13. *Aphanizomenon gracile*. 14. *Anabaena flos-aquae*. 15. *Pseudanabaena catenata* (P. Bourrelly, 1985-1990).

to be taken into account are possibilities of seeding of planktonic organisms from dead water zones or reservoirs.

The densities of different planktonic organisms in Table 4 are theoretical. They do not take into account rates of dilution of water by inputs of affluents, possible floods, successions of species from upstream to downstream, light energy (insolation), and resources of mineral salts for algae. In reality, the density of algae per cubic metre does not go beyond 3.5×10^{10} in the Lot river upstream of Luzech, or 1.86×10^{11} on the Loire at Saint-Laurent des Eaux, in summer. Nevertheless, the table does give a fair idea of the possibilities of development of plankton as a function of transit time.

Table 4. Transit time of waters of the Lot from Entraygues to Aiguillon (310 km) as a function of the flow. Theoretical density of planktonic organisms at Aiguillon, from one individual/m^3 at Entraygues, as a function of the transit time and the growth rates (r).

Flow (m^3/s)	100	50	40	30	20	10
Transit time (days)	11	18	25	32	46	87
Algae/m^3 (r = 0.694)	190	5800	1.6×10^6	4.6×10^6	10^{10}	1.5×10^{18}
Rotifers/m^3 (r = 0.55)	65	960	13,800	2×10^5	8.8×10^7	2.5×10^{14}
Cladocera/m^3 (r = 0.36)	15	90	510	2950	1.6×10^5	2.7×10^9
Copepoda/m^3 (r = 0.14)	3	6	11	22	105	4600

Algae have a high growth rate and rapidly colonize a medium. The fecundity of rotifers is low (10 to 30 eggs per female), but it is compensated for by a very short cycle (1.6 to 3 days, at 20°C, between two generations). Cladocera and Copepoda have a high fecundity but a long cycle (7 to 9 days at 20°C for Cladocera, 21 to 26 days for Copepoda). From algae to crustaceans, the smaller the organism, the shorter the cycle. Algae and rotifers appear to be the pioneer colonizers. They can proliferate or disappear rapidly as a function of environmental conditions. The result is heterogeneity and instability over time in the distribution of plankton in running waters, which is characteristic of pioneer ecosystems (and confirms the important role of seeding).

Some small Cladocera, such as *Bosmina longirostris* or *Chydorus sphaericus*, can also rapidly colonize an environment. This is not true of Copepoda.

From these trends the mechanisms of plankton development in running waters can be better understood. It is in the first place linked to transit time of water in a stream—and thus to the flow and slope. The plankton in a water course appears in the summer in temperate regions and during the dry season in tropical climates.

The Seine and two of its tributaries (the Marne and the Aube) originate on the Langres plateau (the Seine at 476 m altitude). These are plains rivers with a slow current. Phytoplankton appears from May to October, from reaches of drainage order 4 according to the Strahler scale, and culminates at order 8. Zooplankton is characterized by the dominance of rotifers. The diluting effect of the tributaries, which can still be seen at order 3, subsequently diminishes and practically disappears at order 8.

Up to Entraygues, the slope of the Lot river varies from 1.4 to 0.3%. This is the rhithral level and algae in suspension are of benthic origin. The phytoplankton develop downstream of Entraygues (order of drainage 6) for a slope of 0.1 to 0.05%. The density of rotifers is high only from Villeneuve-sur-Lot (km 440).

Plankton found in the Loire was studied from Belleville-sur-Loire (km 537, altitude 136 m) to Chinon (km 796, altitude 33 m). The average slope was 0.04%. Phytoplankton and rotifers were abundant from Belleville-sur-Loire onwards, and their density increased until Saint Laurent des Eaux, where it reached 1.86×10^{11} algae and 1.3×10^{5} rotifers per m^3. Cladocera (690 individuals/m^3) and Copepoda (417 individuals/m^3) are rare. These are for the most part species not closely associated with aquatic vegetation; that is, they are more drifting than planktonic.

If the transit time determines the development of plankton, the biomass and composition of plankton must be variable from one year to another, as a function of the flow. In 1971, the flow of the Lot at Cahors became less than 50 m^3/s only from mid-August onwards (transit time over 310 km, 18 days), and the phytoplankton was poorly developed. In July 1973, the phytoplanktonic biomass was 3.5 times that of July 1972.

On the Loire, a succession of unusual floods disturbed the flows during the summer of 1977. The planktonic biomass was 4 to7 times less than that of the summer of 1978.

At very short time scales—a day, for example—the density and composition of plankton at an identical point differ according to the water flow. On the Lot, the flow varies at a ratio of 1 to 10 during a summer day, depending on the production requirements of hydroelectric stations. At an identical point, masses of water of different qualities pass by, with plankton of density and quality that vary in a single day. Each water mass has its own history.

2. Modelling of phytoplankton development and seasonal successions

Various models have been designed to simulate the dynamics of phytoplankton in a stream, as a function of water transit time, luminosity, temperature, and nutrients (nitrogen, phosphorus, and silica). The model tested on the Lot, downstream of Entraygues, assumes a regularised flow and comprises eight species of algae (diatoms, Chlorophyceae, and Cyanophyceae), each having its own growth characteristics. The variables are those observed on the Lot from June to October, for flows of 10 to 100 m^3/s. The increase in flow induces a reduction of the algal biomass and a modification of the structure of the settlement (Fig. 27). High flows correspond to the dominance of diatoms, the Chlorophyceae biomass being significant only in the lower

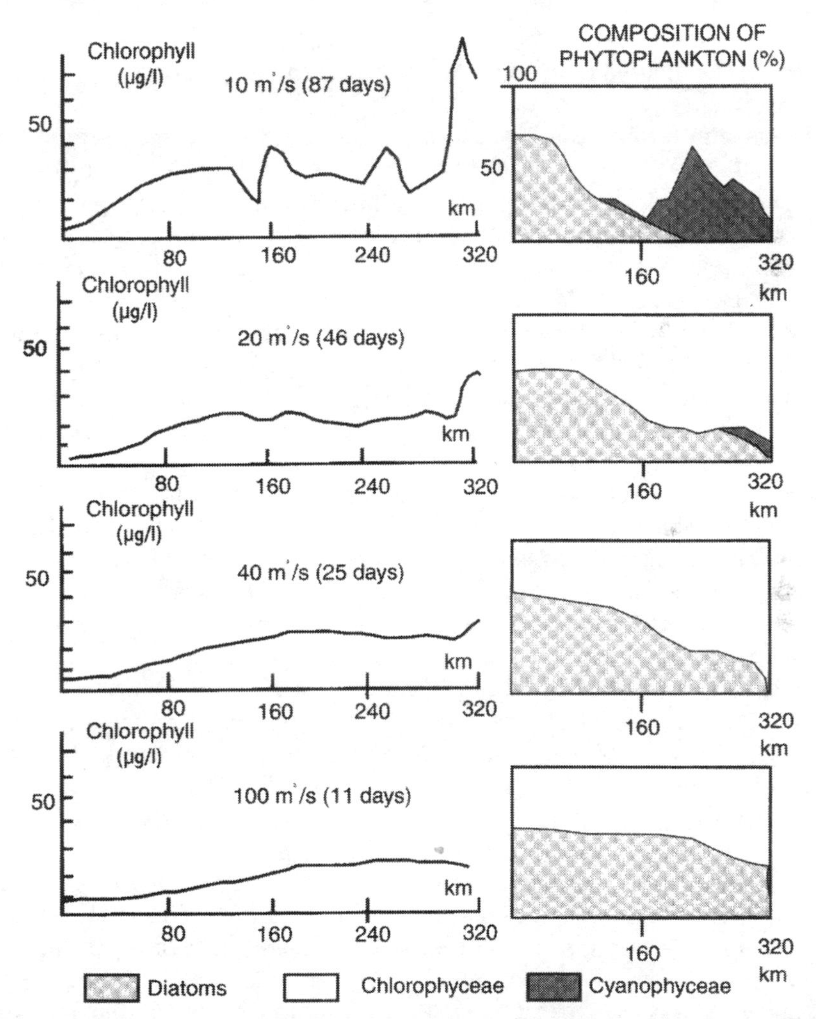

Fig. 27. Evolution of biomass (in µg/l chlorophyll) and of plankton composition in the Lot river in summer (starting date, June 1) as a function of the flow. Simulations were done from observations made on the Lot river and from the growth rates of eight species of algae, as a function of temperature, light, and nutrients, for a regular flow (A. Dauta, 1986).

course. When the flow diminishes, Chlorophyceae develop from the middle course onwards. Cyanophyceae really develop only at flows less than 20 m^3/s (transit time longer than 46 days).

For low flows (transit time longer than 50 days), the upstream-downstream evolution of phytoplankton is marked by a succession of phases of multiplication and decline, in relation to the concentration of nutrients (resulting from their consumption by algae). Cyanophyceae show phases of sudden proliferation—blooms—when nitrates are exhausted (they directly use dissolved nitrogen). Their mortality after a bloom releases nutrients by means of which other species resume development. The simulation of Cyanophyceae blooms required a sub-model of the catastrophic type (in the mathematical sense of the term, discontinuity).

These simulations correspond to different flows in the summer (June to September). In the autumn, they show decline in planktonic biomass and its downstream trajectory, due to light reduction and a drop in temperature. These models can be used to interpret the seasonal successions of plankton in a river.

From December to March, diatoms represent 95% of the planktonic biomass on the middle and lower Lot. However, this biomass is small (660,000 cells/m^3 at most). In April-May, they develop rapidly in cold waters, but with high luminosity, and despite short transit times. Most diatoms have a high rate of growth at low temperatures (optimum growth at 20°C).

From June onward, for high temperatures, high luminosity, and long transit times, the diatoms remain dominant from Entraygues to Cahors, but they yield to Chlorophyceae on the lower part of the Lot (the optimum growth of species used in the model occurs at 30°C). The latter progressively gain on the middle Lot during summer and end up representing 46% of the algal biomass at Cajarc.

At the lowest flows in summer, the successions of phases of multiplication and decline of phytoplankton, for stabilized temperatures and luminosity, correspond to concentrations of nutrients in the water, with blooms of Cyanophyceae (Chapter 14).

In October, phytoplankton in the Lot river is fairly similar to that in the summer, but it is carried further downstream.

The same types of seasonal successions are observed in the Loire. From March onwards, the diatoms develop progressively, from downstream to upstream. The biomass is significant from June onward, with diatoms dominant and Chlorophyceae beginning to develop. In August, at Belleville-sur-Loire (km 537), the biomass is at its highest with diatoms and Chlorophyceae; at Saint-Laurent-des-Eaux (km 791), Chlorophyceae are dominant. The diatoms dominate again in the autumn and persist in winter at low densities.

As for the rotifers, the dominant species evolve from one year to another. A certain number of species first appear downstream of nuclear plants, in the warmer waters, and remain until later in the season.

3. Conclusions

Plankton of running waters is primarily linked to the current itself, which limits its development on courses with steep slopes, probably by mechanical action (turbulence, friction). It is thus characteristic of the potamal of plains rivers and canals.

When the current no longer plays an ecological role by its mechanical action, it plays another role by the intermediary of water transit time. Water transit time is a function of the surface of the submerged section (width and depth of the bed) and of the flow, and it determines the possibilities of plankton development.

For short transit times, diatoms dominate, especially at low temperatures. When the transit time increases, Chlorophyceae begin to develop, in relation to summer low water and high temperatures.

When transit time is longer (about two months or more in the summer, in the Lot), it is no longer a limiting factor. At this point, luminosity, temperature, and nutrient levels become the regulatory factors of algal development. It is in these conditions that eutrophication phenomena may appear, in association with high nutrient concentrations (Chapter 14).

At very low flows in summer, blooms of Cyanophyceae appear. They correspond to long transit times, in the Lot as well as in the Vire river in Normandy. These blooms are chaotic and reflect the time and space heterogeneity of pioneer organisms.

Transit time thus appears to be the primary regulator of phytoplankton. When it is no longer truly limiting, nutrients, nitrates, and phosphates in turn become regulatory factors. When these are in excess, temperature and light amplify algal development. This major role of transit time constitutes the essential difference between plankton of running waters and that of stagnant waters.

The problem presented by running waters is to determine whether competition between species of phytoplankton and the influence of browsing by zooplankton have an effect on their density, as they do in stagnant waters.

Simulations were conducted from characteristics of waters from the Lot, with one, three, and eight species of algae. The response was linked to conditions of growth of each species when it is alone. From three species, the model of development was very similar to that realized with eight species and introduced simply seasonal and upstream-downstream successions. Competition did not really appear to be significant.

As for zooplankton, the transit times observed on the Lot do not allow sufficient development for browsing to have a regulatory effect on phytoplankton. However, regressions of algal density were observed during proliferation of rotifers and regulation of phytoplankton by protozoa and rotifers on the Seine.

A final important phenomenon is the consequence of algal photosynthesis. On a rhithral, without true phytoplankton, the oxygen concentration is stable, close to saturation. Turbulence favours exchanges at the water-atmosphere interface. When turbulence phenomena diminish and photosynthesis is significant, there is a daily cycle of oxygen concentration with a supersaturation during the day. However, the degradation of Cyanophyceae that follows a bloom leads to consumption of oxygen and a phase of sub-saturation, even of passing anoxia.

8

Fish of Running Waters

Many species of fish in running waters also colonize stagnant waters and are sedentary. Some species migrate to the sea to reproduce, for example the eel (catadromous migration). Others live in the sea and come to fresh water to reproduce, as with salmon or shad (anadromous migration).

Out of more than 60 fish species found in France, 22 are Cyprinidae (e.g, carp, tench, bream), and about 10 are Salmonidae (e.g., common trout, salmon).

1. Swimming and the water current

The ability of a species to sustain itself in running water and resist being swept away depends primarily on its swimming speed. Swimming speed varies greatly from one species to another and determines the minimum speed at which it can survive in the current (Table 5).

Table 5. Speed limit of some fish species for maintenance in the current (Kreitmann, 1932)

	Speed limit (m/s)
Salmon	8.00
Common trout	4.40
Chub	2.70
Barbel	2.40
Bream	0.60
Tench	0.50
Pike	0.45
Carp	0.40

Apart from the swimming speed, the duration for which a fish is able to swim, as well as the shape of its body, determine its resistance to the current. The minimal resistance opposed to the current by a fish corresponds to a body shape in which the largest cross-section is located at a little more than one third of its total length (torpedo shape). When the cross-section of the body is oval—as in trout (*Salmo trutta*) or chub (*Leuciscus cephalus*), for example—the musculature is highly developed and allows strong resistance to the current (Fig. 28). Conversely, laterally

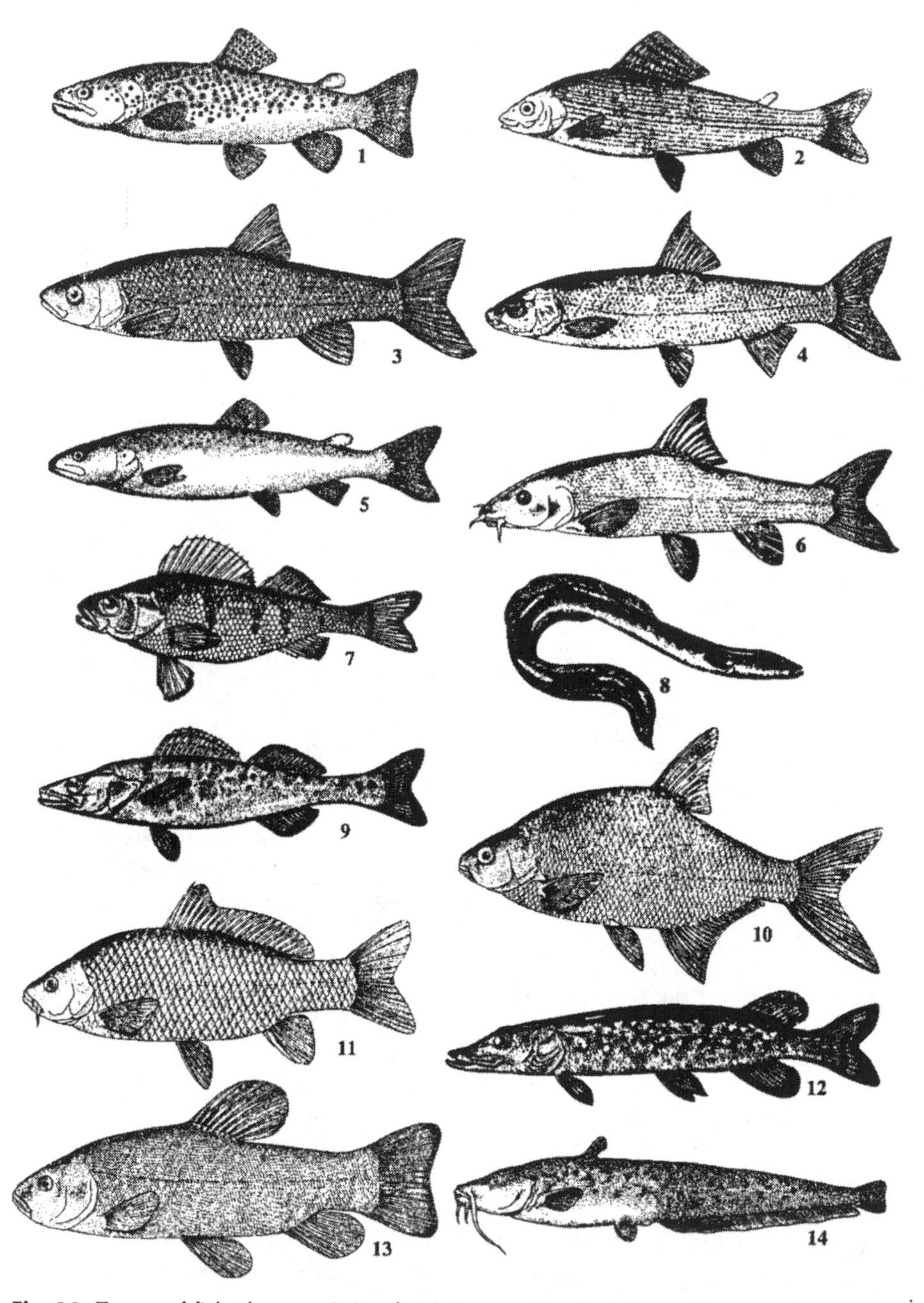

Fig. 28. Types of fish characteristic of running waters. 1. *Salmo trutta* (common trout). 2. *Thymallus thymallus* (grayling). 3. *Leuciscus cephalus* (chub). 4. *Chondrostoma nasus* (nase). 5. *Hucho hucho* (Danube salmon). 6. *Barbus barbus* (barbel). 7. *Perca fluviatilis* (perch). 8. *Anguilla anguilla* (eel). 9. *Stizostedion lucioperca* (pikeperch). 10. *Abramis brama* (bream). 11. *Cyprinus carpio* (carp). 12. *Esox lucius* (pike). 13. *Tinca tinca* (tench). 14. *Silurus glanis* (Wels catfish) (C.J. Spillman, 1961).

flat fish such as bream (*Abramis brama*) have reduced musculature and do not effectively resist the current.

The potential duration of swimming effort also varies according to the species. The swimming effort of the tench (*Tinca tinca*) is low. That of the carp (*Cyprinus carpio*) and bream is more vigorous but not sustained for long. In barbel (*Barbus barbus*) as well as chub, the resistance capacity diminishes when the effort is prolonged. Swimming effort is prolonged in common trout and salmon (*Salmo salar*) but involves a significant energy expenditure.

Life in the full current, which requires a constant swimming effort, therefore can only be temporary. All species remain habitually close to the bottom or the banks, behind obstacles, and sheltered from the current. They swim in the current only to feed. This behaviour of fish is analogous to that of invertebrates (Chapter 4). For trout, the substrate itself serves as a shelter. The gravel and pebble bottom constitutes shelter sufficient for small juvenile trout. For adults, only coarse beds of rocks and stones, in the deeper zones, offer optimal shelter.

2. Distribution of fish on a longitudinal profile

The varying swimming speed limit of different fish species results in an upstream-downstream succession of these species on a longitudinal profile of the river, as a function of the slope. Species that swim rapidly, such as trout, colonize the upper course of rivers, while carp, tench, bream, and pike (*Esox lucius*) are confined to the lower course, to calmer waters, on the bed or in the vegetation.

M. Huet (1949), from observations on upstream-downstream successions, suggested a typology of piscicultural zones as a function of the slope and width of the bed (Table 6, Fig. 29). This involves a mapping of the position of successive piscicultural communities along the profile and corresponds to Strahler's order of drainage (Chapter 2). Indeed, the succession of communities depends not only on the slope and width of the bed (or order of drainage), but also on summer temperature, which regularly increases from upstream to downstream. Huet distinguished four successive zones, dominated by trout, grayling (*Thymallus thymallus*), barbel, and bream.

The trout zone is characterized by slopes greater than 4.5‰ and a bed width from minus 1 m to 100 m. This is the upper course of the river, by the slope, with a low drainage order number.

The grayling zone corresponds to slopes from 1‰ (for a width of 100 m) to 4.5‰ (for a width less than 1 m). Species richness is higher than in the trout zone with, apart from grayling and trout, cyprinids of rapid water such as chub, barbel, and nase (*Chondrostoma nasus*).

Table 6. Huet's piscicultural zonation (1949)

Rivers of 1st category, dominated by Salmonidae		Rivers of 2nd category, dominated by Cyprinidae	
Trout zone	Grayling zone	Barbel zone	Bream zone
Trout	Trout	Barbel	Bream
	Grayling	Chub	Carp
	Chub	Nase	Tench
	Barbel	Grayling	Roach
	Nase	Roach	Pike
	Danube salmon	Perch	Perch
		Pikeperch	Pikeperch
		Eel	Eel
		Danube salmon	

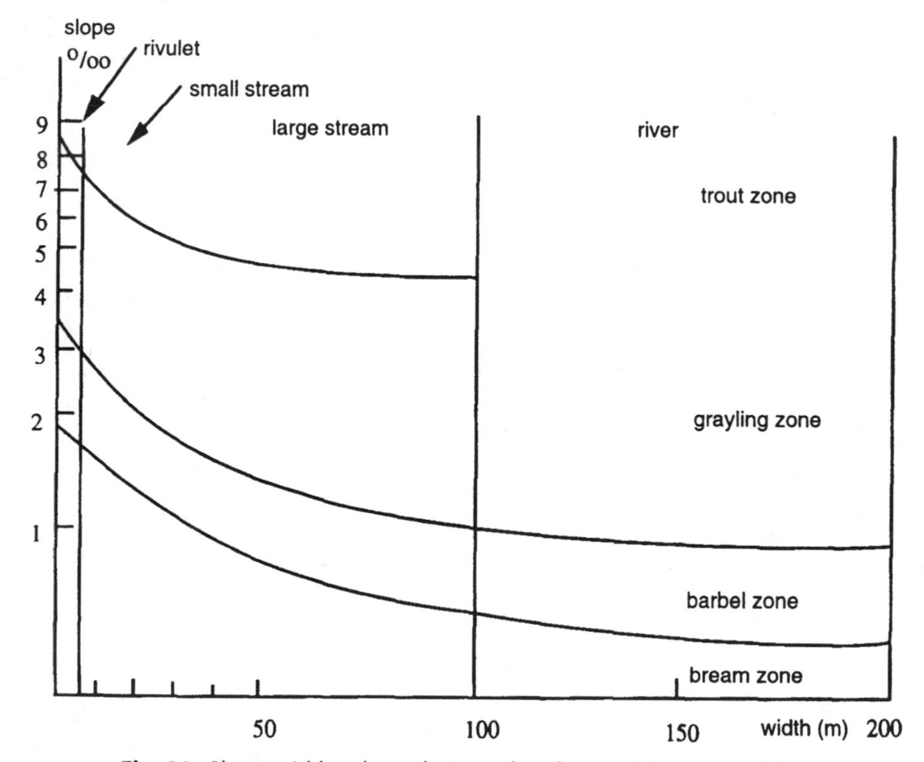

Fig. 29. Slope-width rule and piscicultural zone (M. Huet, 1949)

Trout and grayling zones constitute what is referred to in fishery laws as waters of the first category, with salmonids dominating. These are cool waters, with a summer temperature that does not exceed 20 to 22°C. Salmonids reproduce from the end of autumn to the end of winter—November to January for common trout and brook trout (*Salvelinus fontinalis*), and March to April for grayling (which is why fishing is prohibited in rivers of the first category during the reproductive period). Egg-laying occurs on the gravel substrates, and the problem of spawning is critical in rivers dominated by Salmonidae. The granulometry of the substrate is that of a rhithral.

Barbel and bream zones are also dominated by Cyprinidae (streams of the so-called second category). The barbel zone corresponds to slopes of 0.2 to 1‰ (for a width of 100 m) and 1.5 to 3.5‰ (for a width of 1 m). Grayling still remains, but barbel, chub, and nase are dominant, along with carnivorous fish such as perch (*Perca fluviatilis*), pikeperch (*Stizostedion lucioperca*), and eel (*Anguilla anguilla*).

The bream zone is in the calmest waters, on lower courses of rivers, with high summer temperatures. Bream is accompanied by carp, tench, roach (*Rutilus rutilus*), bleak (*Alburnus alburnus*), and the carnivorous pike, pikeperch, black bass (*Micropterus salmoides*), perch, and eel.

The summer temperature rises to more than 22°C and the waters are those of a potamal. The reproductive period lasts from mid-March for the most precocious species (pike, perch, dace (*Leuciscus leuciscus*), pikeperch) to the end of June for the latest species (tench, carp, bream, chub, nase). Fishing is therefore prohibited during spring, the reproductive period.

This slope-width rule, which corresponds also to the summer temperature, is valid for temperate Western European rivers. In rivers of Central Europe, the slopes of the same zones are higher for the equivalent widths: 6 to more than 10‰ for the trout zone, 2.5 to 5‰ for the grayling zone, 0.5 to 2‰ for the barbel zone. In the rivers of Normandy, in waters that are cool in summer (9 to 11°C), the trout zone begins on slopes of just 1‰ for widths that do not exceed 50 m, but with significant flows. This happens because the speed of the current is a function of the slope and flow for an equivalent bed depth. It also emphasizes the role of temperature in the distribution of fish species.

In the Garonne basin, two distinct groups appear in the bream zone: one consisting of carp, tench, perch, pikeperch, and pike, and the other, further downstream, of eel, bream, pumpkinseed sunfish (*Lepomis gibbosus*), pikeperch, and black bullhead (*Ictalurus melas*).

3. Migration of fish

Many species of fish are sedentary. Others migrate during certain times of the year for reproduction or for feeding. Barbel, during the reproductive period, assemble on the banks and swim upstream in search of gravel or rocky bottoms on which they can lay eggs.

There are sedentary populations of common trout—colonizing habitats that are favourable for reproduction and growth—as well as semi-migratory populations. The latter move from a habitat suitable for their growth to one suitable for reproduction and the growth of juveniles, generally in tributaries. The one- or two-year-old juveniles in turn swim back downstream to the main course. In addition, there is a truly migratory population of common trout—sea trout—on the coastal rivers and downstream of the large rivers of Western Europe, from the White Sea to southern Spain. It lives in the sea and returns to the streams to reproduce (anadromous migration). The adults swim upstream into fresh water from May to January (with a slack period in August-September). The new generation returns to the sea from the end of November to mid-May. The fish reach their feeding sites, where they grow to adulthood (in the English Channel and the North Sea for the populations of Normandy-Picardy).

Shad (*Alosa alosa* and *A. falax*) swim upstream in May-June. The young swim downstream in autumn to feed and grow to adulthood in the sea.

The adult salmon reaches the river mouth late in the winter and up to the end of spring. It is there that it reaches sexual maturity before swimming upstream again to its spawning grounds, in the same zones as the common trout. After egg-laying, the parental generation returns to the sea, but many salmon are exhausted and fail to make it. The eggs hatch between February and May, depending on the altitude and the latitude. The young salmon lead the lives of trout for 2 to 3 years. They then take on a silver coating (smolt) and, during a spring high-water period, return to the sea, where they remain for 1 to 4 years.

Eel migration is the reverse of that of salmon or sea trout (catadromous migration). The eel grows in fresh water (rivers or lakes) or brackish water and remains there for 4 to 10 years, at the end of which during autumn the adult eel descends the rivers to the sea. After a voyage of 4000 to 7000 km, depending on its origin, it reaches the Sargasso Sea, where it lays eggs. The young (leptocephales) drift with the Gulf Stream and arrive in Europe at the end of 3 years. They metamorphose into elvers and colonize brackish waters or swim upstream into the rivers.

4. Geographical distribution of fish

The geographical distribution of fish is not uniform, even over a small area such as France. The relative isolation of watersheds does not facilitate faunal exchanges. For example, Huet defined a grayling zone succeeding a trout zone. The grayling is in fact indigenous only to Central and Northern Europe, northwestern France, and the upper basins of the Rhone and the Loire.

To understand the distribution of fish in Europe, the geological history of this region during the Tertiary Era, when the present species appeared, must be recalled. The Pyrenees emerged definitively during the Eocene Epoch, and the most significant folding of the Alps took place during the

Oligocene period. These ranges became barriers to the exchange of fauna between middle Europe and the Mediterranean regions.

Cypriniform fish, originating from Central Eurasia to the northern Himalayas, expanded towards Asia and Europe during the Tertiary period. In the Pliocene period, European ichthyological fauna was rich and relatively homogeneous, at least north of the Alpine and Pyrenean barriers.

The Quaternary glaciations decimated the most thermophilous fauna, because the mountain chains, oriented east-west (unlike in North America, where they run north-south), and the Mediterranean Sea limited the retreat of species towards the south. Between the Scandinavian ice cap and the Alpine and Pyrenean glaciers, a cold-water stenothermal fauna survived, notably amphihaline migratory species such as the eel and Salmonidae.

Some thermophilous species survived south of the Alps and the Pyrenees, but the true refuge of ichthyological fauna was the middle and lower Danube valley, a basin vast enough to shelter a large diversity of species.

Since the end of the glaciations, European waters were recolonized mostly from the Danube basin, where more than 100 species are presently known. Species richness diminishes from east to west: there are 60 species in the Rhine basin, 56 in the Rhone basin, 50 in the Loire basin, 45 in the Seine basin, and 47 in the Garonne basin (including estuary species and introduced species).

Several factors favoured the dissemination of species: capture of water courses in the apical parts of hydrographic networks, confluences of courses downstream, and canals linking large river basins.

Perch, indigenous to northern France, colonized the Garonne and Herault basins only during the 19th century. It was introduced to Spain, like dace and pike. The nase (*Chondrostoma nasus*) extended from the northeast to the basins of the Seine, Loire, and Rhone at the end of the 19th century, colonizing the grayling zone. The Wels catfish (*Silurus glanis*), the largest carnivore of European waters, moved into the Rhine and the Doubs from the Danube. The pikeperch invaded the Rhine basin during the 19th century, then, via canals, the Doubs, Saone, and Rhone rivers during the 20th century.

The northward spread of the southern species of Europe was relatively limited. The toxosome or French nase (*Chondrostoma toxostoma*), a relative of the nase, lived in Italy, Spain, and the southwestern and southeastern parts of France. Its northward extension was limited to the Loire and Allier basins, in the Massif Central, and the Saone. The nase seems to have competed with it and eliminated it. The Mediterranean barbel (*Barbus meridionalis*) colonized the Mediterranean water courses (Var, Argens, Herault) and the middle basin of the Rhone.

To the recent migrations must be added introductions from one basin to another—e.g., grayling, pikeperch, Danube salmon, Wels catfish. The pikeperch was introduced to the basins of the Seine, Loire, and Garonne, as was the Wels catfish. Since the 1980s, a veritable demographic explosion of Wels catfish has been observed in these basins, as well as in the Saone and its tributaries. The Danube salmon was first acclimatized in the Rhine basin, then in some Alpine streams, from the Danube. It colonized the upper part of the barbel zone. The oldest introduction in Western Europe is that of the carp, by the Romans. The species was widely disseminated in the Middle Ages by religious communities.

It was from the second half of the 19th century that North American fish species were first introduced to Europe. Some acclimatizations were successful: brook trout on the Atlantic side, from North America, and rainbow trout (*Oncorhynchus mykiss*) on the Pacific side. They colonized the trout zone and rainbow trout extended to the grayling zone. Black bass was introduced to Germany and the Netherlands from 1863. It is a carnivore that colonizes barbel and bream zones. The species *Micropterus salmoides* is found in all the French river basins, while *M. dolomieu* is limited to Belgium and northwestern France.

Other acclimatizations have failed, as with sun perch and catfish, both North American fish that were introduced into European barbel and bream zones. The catfish is an omnivore that tends to eliminate other Cyprinidae, and proliferation of the sun perch resulted in its dwarfism. Black bass is its natural predator.

The problem with acclimatization is the following: Is there any ecological niche that is unoccupied, an unexploited source of food? If there is, acclimatization of new species has the advantage of increasing fish resources. Salmonidae and black bass are good examples. The elimination of indigenous Cyprinidae by catfish, which had no natural predators before the introduction of black bass, made it a pest fish. The same was true of sun perch, the success of which led to proliferation and dwarfism.

5. Conclusions

Swimming speed and duration of swimming effort, as well as temperature, determine the distribution of fish on the longitudinal profile of a river. The zonation of fish, as a function of the slope and width of the water course, reflects the swimming capacity of the species and the temperature.

Swimming can only be temporary, limited to movement from one place to another in search of food. It is not a necessity. Nearly all running-water species also colonize lakes. Survival in a water course is ensured only by the presence of resting areas—shelters, vegetation, bank edges, deep-water zones with a slow current, and stones and large pebbles of the substrate.

Swimming effort involves significant energy expenditure and hence slows the organism's growth. At the same age, a river trout is smaller than a lake trout.

Resistance to the current is limited by its speed as well as its duration. Flash floods carry away some fish, especially those in the juvenile stages.

Ecological factors are not by themselves sufficient to explain the distribution of fish. Distribution results from the history of the population since the Tertiary Era and the original isolation of river basins. The distribution of European fish has been modified since the end of the 19th century by human intervention, including construction of canals linking the large river basins and introduction of species from one basin to another or from one continent to another.

9

Temperature, Biological Cycles and Distribution of Organisms

Temperature and current are the most important ecological factors in running waters (Chapter 2). Temperature varies regularly along the longitudinal profile of a water course in relation to atmospheric temperature. It determines the potential for development and the duration of the biological cycle of each species. At each point on a longitudinal profile, temperature is linked to altitude, distance from the source, hydrological regime, and season.

A species can survive between two temperature limits. Between these limits, temperature has an effect on its metabolism, duration of its biological cycle, its survival term, and its rate of reproduction.

1. Temperature and development of organisms

1.1. Temperature thresholds and temperature of maximum activity

The ecological importance of heat arises from the general dependence of chemical reactions on heat: heat is the essential regulator of cellular activity.

Cellular activity can occur only between two limits—the minimum and maximum temperature thresholds. Between these two limits, the activity curve (which can be measured by the respiratory metabolism) as a function of temperature is bell-shaped; its peak corresponds to the maximal activity temperature. This curve is asymmetrical: between the temperature of maximal activity and the upper temperature threshold, the activity diminishes rapidly, probably because of lesions on the protoplasm and fracturing of the coordination between the various physiological processes. These lead to the death of the individual: the maximal temperature threshold most often corresponds to the lethal temperature.

Between the minimal temperature threshold and the temperature of maximal activity, the metabolism-temperature curve can be compared to that of a chemical reaction—an exponential branch approximately following the Arrhenius law: temperature has a multiplying effect on metabolism (Q_{10} measures the multiplying effect for a temperature increase of 10°C).

In these conditions, the biological cycle and the altitudinal distribution of flora and fauna of running waters should be explained from the metabolism-temperature curve of each species. The difficulty lies in the interpretation of results. Because of the multiplicity of physiological processes that occur simultaneously in an organism, metabolism cannot be compared to a simple chemical reaction. Factors other than temperature also intervene (e.g., oxygen level, current, nature of the substrate).

It has been shown, for some Crustaceae and larvae of insects, that the difference in respiratory metabolism between 5 and 25°C was a specific constant independent of experimental conditions other than temperature. A classification of similar species based on the difference in oxygen consumption between these extreme temperatures conforms to their upstream-downstream succession on a longitudinal stream profile and to the number of annual generations (Table 7).

Table 7. Range of oxygen consumption (mm^3 O_2/mg dry wt/h) between 5 and 25°C and upstream-downstream distribution of Trichoptera Hydropsychidae and Amphipoda Gammaridae (C. Roux, 1989)

Upstream	*Hydropsyche dinarica* monovoltin (997)	*Gammarus fossarum* (1225)
↓	*Hydropsyche siltalai* monovoltin (1102)	*Gammarus pulex* (2137)
	Hydropsyche pellucidula mono- or bivoltin (1308)	
	Hydropsyche contubernalis bivoltin (1771)	
Downstream	*Hydropsyche modesta* bivoltin (2162)	

Between temperatures of 5 and 25°C, four types of metabolism-temperature curves can be identified (Fig. 30):

(A) a straight line with a steep slope corresponding to an adjustment of the metabolism to the temperature;

(B) a curve approaching an exponential function, with a regulation zone followed by an inflexion point and an adjustment zone;

(C) a curve having a regulation zone (a plateau) in the range of average temperatures, preceded and followed by an adjustment zone;

(D) a curve characterized by the low variation of the metabolism between extreme temperatures, with a plateau at high temperatures.

The first three curves pertain to species of the middle and lower course (end of the rhithral and potamal) that are tolerant of wide temperature

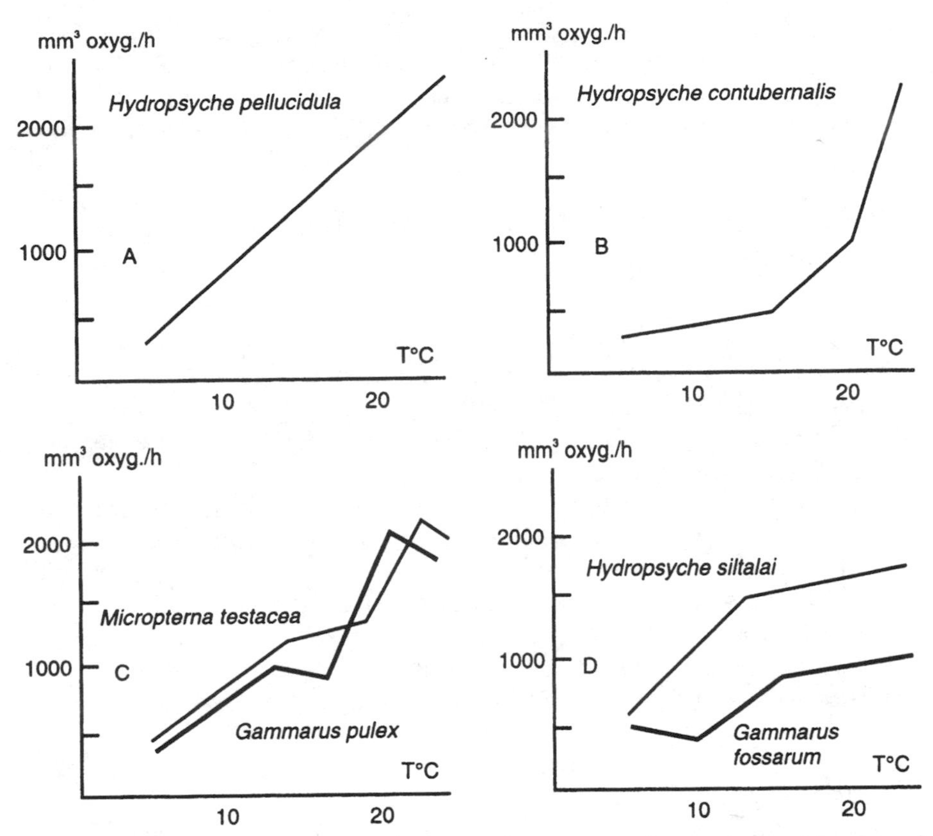

Fig. 30. Oxygen consumption as a function of temperature in certain Amphipoda and Trichoptera (C. Roux, 1989)

variations (*eurythermic* species) or develop only at high temperatures (*stenothermic* species of warm waters), with several annual generations (*polyvoltin* species). The fourth type of curve corresponds to *stenothermic* species of cold waters, with a single annual generation or a biological cycle extending over more than a year (mono- or semi-voltin species).

The temperature range corresponding to a possible plateau is useful to know. It may explain the replacement of taxonomically similar species over the course of a year. In Britain, two Trichoptera Hydropsychidae—*Diplectrona felix* and *Hydropsyche fulvipes*—have an identical lethal temperature, 28°C. The plateau of the metabolism-temperature curve is located at higher temperatures in *H. fulvipes* than in *D. felix*. When the summer temperatures of water courses rise beyond 15°C, *H. fulvipes* replaces *D. felix*.

The four types of metabolism-temperature curves show the limitations of the usefulness of Q_{10}. Q_{10} implies a constant adjustment of the metabolism to temperature and ignores the regulation zones and the plateau. Between

5 and 25°C, with regulation zones, very different Q_{10} may be obtained for a single species, depending on the adjustment zones. In Trichoptera *Limnephilus rhombicus*, for example, Q_{10} is 0.8 between 5 and 10°C, 3.99 between 10 and 15°C, 1.28 between 15 and 20°C, and 1.06 between 20 and 25°C.

1.2. Lethal temperatures, limits of indefinite survival and population growth rates

The metabolism-temperature curve has shown that, beyond the maximum temperature of activity, there is a zone that very rapidly becomes lethal. Thus, a zone that is lethal in 24 h can be defined.

The minimal and maximal temperatures of activity may vary as a function of acclimatization temperature, which is obtained by slow increase of the original temperature (1°C per day).

In Crucian carp acclimatized at 0°C, the temperature of maximal activity is 26°C. It is 33°C for acclimatization at 15°C. For acclimatization beyond 33°C, the maximal temperature no longer exceeds 40°C, but the minimum temperature threshold is no higher than 0°C (respectively 9 and 17°C for acclimatization at 38 and 40°C). The same phenomenon is observed in catfish. These two types of fish are largely eurythermic, but the temperature thresholds vary according to the acclimatization temperature.

Inversely, in Salmonidae, which are stenothermic fish of cold waters, temperature thresholds practically do not vary as a function of the acclimatization temperature.

Apart from metabolism, the difference between eurythermic and stenothermic species may thus result from their potential to acclimatize themselves: temperature thresholds are thought to be variable in eurythermic species, but not so in stenothermic species.

The temperature that is lethal in 24 h finds expression on the population scale by sudden destruction. However, within the range of temperatures of activity, temperature also influences the life span of a population in the medium or long term. This phenomenon has been studied particularly in Turbellaria (Fig. 31).

At 20-21°C, the mean survival time of *Dugesia gonocephala* is longer than 400 d. This allows a population to survive or increase by means of succeeding generations. From 22.5°C onward, the average survival time decreases to 176 d, and it is no more than 2 d at 27.5°C. From 22.5°C onwards, there is a resistance zone, in which the increasing mortality rate ultimately decimates the population. Below 22°C, there is a tolerance zone, in which survival time is indefinite in relation to temperature.

The limit of the tolerance zone of three succeeding species of Turbellaria normally on a longitudinal profile of the water course is 12°C for *Crenobia alpina*, 16°C for *Polycelis felina*, and 21°C for *Dugesia gonocephala*. In the tolerance zone, fish may face high temperatures, but

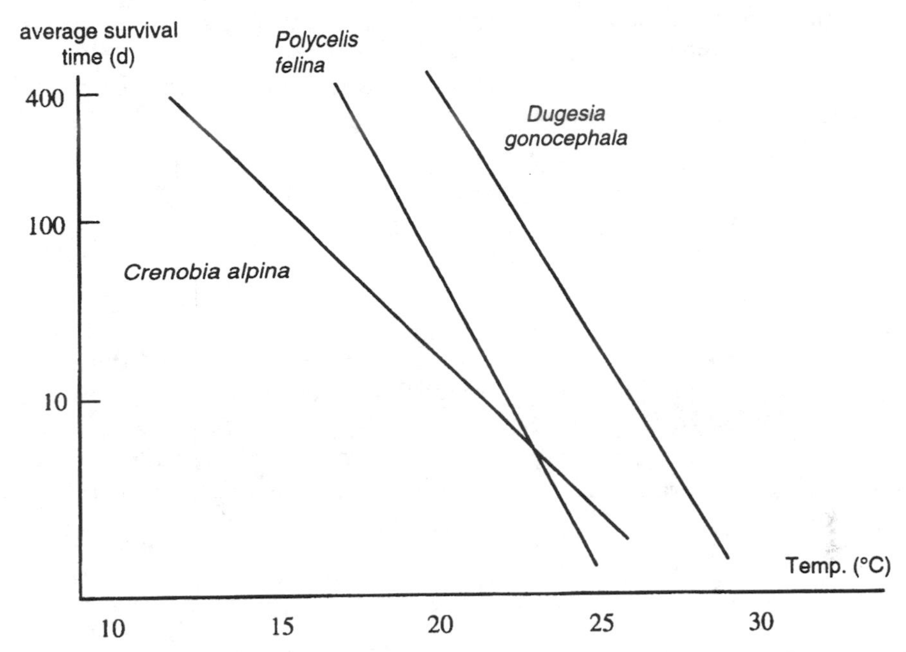

Fig. 31. Average survival time of three species of Turbellaria as a function of temperature (E. Pattee, 1958)

only for a limited time. *Crenobia alpina*, the most stenothermic of the three species, tolerates up to 25°C for a few hours of the day in small streams at high altitude, in which the temperature decreases significantly during the night. Conversely, in streams of the piedmont zone, the fish colonizes only the springs.

Temperatures of the activity zone influence not only the mortality rate in the medium and long term, but also the birth rate—the number of eggs, percentage of hatching, periodicity of the egg-laying season, duration of a generation, and substitution of asexual reproduction for sexual reproduction in Turbellaria.

The growth rate of a population, r, expresses the difference between birth and mortality. $dN/dt = rN$ (N being the size of the population) is therefore a function of the temperature.

Up to 8°C, the growth rate of *Crenobia alpina* is higher than that of *Polycelis felina*. Beyond that temperature, the latter species leads. From 15°C onward, the growth rate of *Dugesia gonocephala* is the highest.

1.3. Temperature and time of development

From the metabolism-temperature curve it can be deduced that the growth rate of a poikilothermic species, and therefore the duration of its development, is a function of the temperature. This applies equally to embryonic development and growth to adulthood.

Within its own temperature limits, a species can be characterized by the following relationship:

Constant of development (degree-days) = (effective degree temperature) × D

where effective temperature is the number of degrees centigrade above the minimal temperature threshold, and D is the duration of development in days.

This development constant shows the total quantity of heat needed for the growth of a species. In a water course, where the temperature varies with the season and time of day, the number of degree-days is obtained by adding up the average daily temperatures. This number depends on altitude and hydrological regime (Fig. 32).

The effective temperature is easy to evaluate for species found at very high altitudes, where it is close to 0°C. The number of degree-days is thus

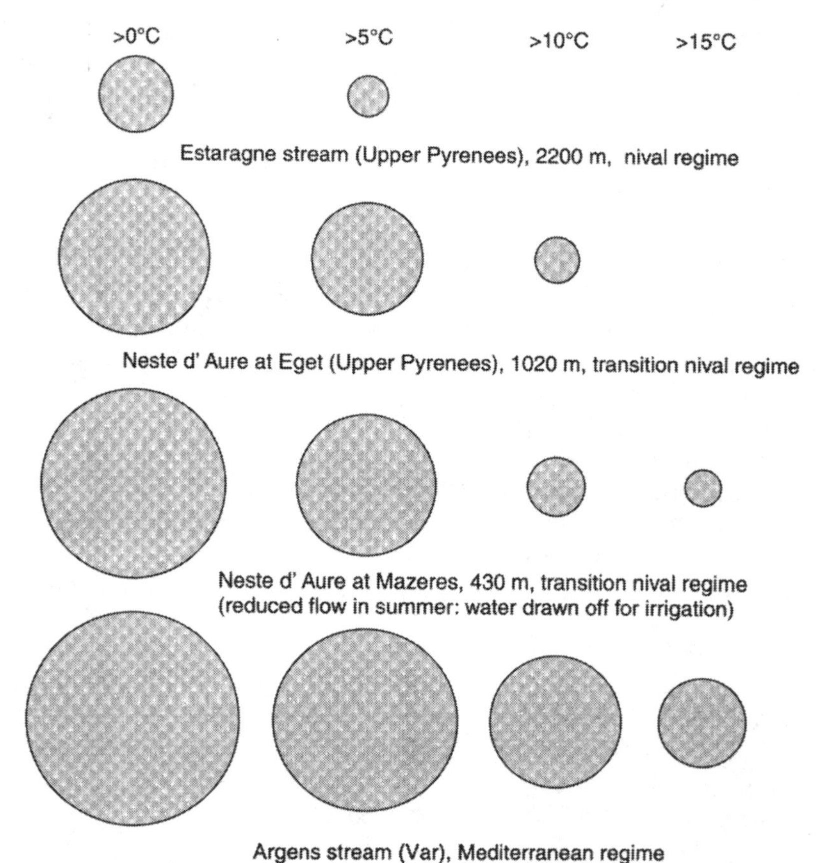

Fig. 32. Representation of number of degree-days above 0, 5, 10, and 15°C, as a function of altitude and water regime

equivalent to that of the torrent, and from this the development constant is easily deduced.

If, instead of simply considering the total number of degree-days, the development is followed over a year, the intervention of factors other than effective temperature is observed, such as duration of the snow-bound phase, high water, or summer temperatures above the upper temperature threshold.

In winter, when a layer of snow covers the bed of a stream, it prevents the penetration of light and consequently photosynthetic processes. The result is a halt in the development of larval stages that feed on algae of the periphyton.

Increase in weight must be distinguished from the stages of development (moults, acquisition of imaginal structures in insect larvae, metamorphosis, sexual maturity). These stages may progress at rhythms different from those of weight increase, at a given temperature.

In the common frog (*Rana temporaria*), the body length at metamorphosis is 11 mm when the tadpole develops at 22°C, and 17 mm when it develops at 10°C.

The larvae at the last stage of *Baetis rhodani* and *Rhitrogena semicolorata* (Ephemeroptera) are smaller in the low-altitude streams of the Atlantic Pyrenees than in the colder waters of Great Britain. In insects that have several annual generations—polyvoltins—the body is longer in larvae of the winter generation than in those of the summer generation. In the Estaragne stream (Upper Pyrenees), the duration of development decreases from upstream to downstream, as does the weight of larvae of the final stage. The temperature increase accelerates the acquisition of imaginal structures more than it does linear and weight growth for most species. However, smaller adults in higher rather than lower altitudes have been observed, in Trichoptera Limnophilidae, for example.

In species of fish that have high longevity and continuous growth, size at sexual maturity diminishes with altitude. The common trout reaches sexual maturity at around 3–4 years and at a length of 12–15 cm at 2200 m altitude in the Pyrenees. In a single river, the length at 3 years varies from 21 cm at 1300 m to 13 cm at 2000 m. In rivers in the plains, trout reach sexual maturity at 2–3 years for lengths of 23 to 27 cm. The legal size for fishing reflects the growth of trout as a function of the temperature (16–18 cm in the mountains, 23–25 cm in the plains).

The problem of growth is complicated by the role of temperature on feeding and duration of digestion. Common trout may totally suspend feeding in the winter. They daily absorb 2% of their weight at 5°C, 5% at 10°C, and 7% at 15°C. Moreover, a significant fraction of received energy is used to resist the current. At equivalent altitude and temperatures, trout are smaller in torrents than in lakes.

As with metabolism-temperature curves, the development constant does not contribute a definitive response to development as a function of temperature, but only certain elements of information.

2. Biological cycles: quiescence, diapause, mono- and polyvoltinism

In the biological cycle of a species, quiescence is a simple halt in development, immediately reversible and determined by the appearance of unfavourable conditions—essentially of temperature in the aquatic environment. Diapause is a phase of slowed-down life that is obligatory in nature, genetically determined and independent of unfavourable conditions.

Fish, which grow constantly, have simple quiescence phases during the winter and can temporarily cease feeding. However, in Protists and Invertebrates, it is more difficult to recognize whether a phase is diapause or quiescence. When a species has only one annual biological cycle (monovoltin species) and individuals disappear over a year's time independently of the prevalence of unfavourable conditions, a diapause can be said to exist.

2.1. Biological cycle of species with diapause

Diapause may appear at various stages of development, but it is most frequent during the egg stage. The minimal temperature threshold determines the beginning of development. Development therefore begins progressively later with increasing altitude. In *Simulium timondavidi* (Diptera Simuliidae) in Corsica, development begins at the end of February and terminates at the beginning of June below 500 m, with a summer and autumn diapause. At 1700 m altitude, larval development occurs from July to October, diapause occurring in winter (Fig. 33). The fact that diapause can occur either in winter or in summer shows clearly that it is independent of temperature conditions. The presence of larva on the longitudinal profile of the river is temporary, and the altitudinal limit is determined by a number of degree-days of the water above the temperature threshold, equivalent to the development constant.

This type of cycle with a diapause is found in many species of Diptera, notably in the Simuliidae, Blephariceridae, Hydracarids, and Plecoptera Capniidae.

2.2. Biological cycles of species with quiescence

The resting phase in species with quiescence occurs when the temperature falls below the minimal temperature threshold, or even above the maximal temperature threshold when that temperature is not lethal.

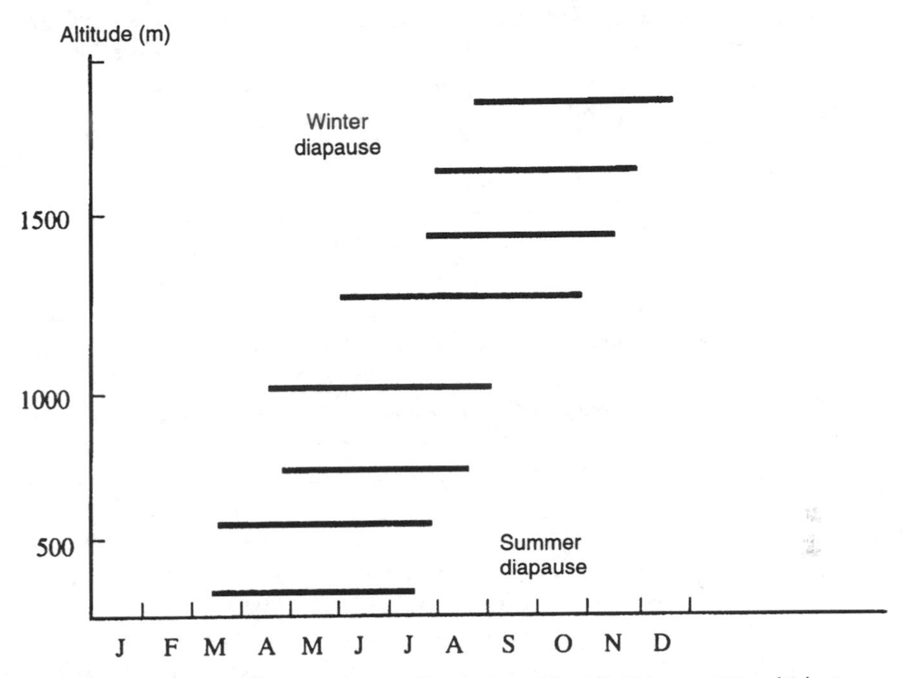

Fig. 33. Larval development of *Simulium timondavidi* (Diptera Dimuliidae) on a Corsican river as a function of altitude (J. Giudicelli, 1969)

Turbellaria, a number of Oligochaetes, Molluscs, Crustacea Amphipoda and Isopoda, and insects such as Coleoptera, with an entirely aquatic life and surviving beyond reproduction, have no significant cyclical variations in their populations. The development constant, as a function of the cumulative number of degree-days of the water, indicates the approximate duration of one generation.

The life of insects that become flying adults ends with the reproductive period. In waters at low altitude, temperate or warm, there are numerous polyvoltin species—with several generations during a year. They are characterized by a low development constant.

In the Lissuraga stream (Atlantic Pyrenees), 10 species of Ephemeroptera out of 17 are polyvoltins, partly or totally. Except for *Baetis muticus* (three generations), they are bivoltin—sometimes partly (*Epeorus torrentium, Caenis moesta*).

Species that are bivoltin in the Lissuraga are monovoltin in colder waters, in Britain and in Central Europe—for example, *Baetis scambus*. Similarly, the Trichoptera *Hydropsyche pellucidula* is bivoltin below 500 m altitude in the Pyrenees or the upper French Rhone (summer temperatures of 16 to 18°C) and monovoltin at higher altitudes in the Pyrenees, in Britain, or in Poland (summer temperatures of 10 to 17°C).

In Central Europe, 70% of Ephemeroptera are monovoltin, 25% are bivoltin, and 5% have a generation spread over two years (semi-voltin) or three years. In the Atlantic Pyrenees, 35% of species are monovoltin, 60% polyvoltin, and 5% semi-voltin.

Polyvoltinism characterizes waters with a high annual number of degree-days. In the Lissuraga stream, the number of degree-days above 0°C is 4470 (it is 3400 to 3600 in comparable English streams). It is an additional 145 to 175 above 15°C (10 to 66 in Britain).

As the number of degree-days decreases, depending on the altitude or latitude, monovoltinism (with diapause or quiescence) becomes dominant; the biological cycle then extends longer than one year and even over several years, either with the same species or with different species.

Baetis alpinus, for example (development constant 1290), is polyvoltin at low altitude (two or perhaps three annual generations, according to different authors). It becomes monovoltin at higher altitudes, then semi-voltin at 1850 in the Central Pyrenees. Its cycle is spread over 3 to 4 years beyond 2100 m.

The cycle of Trichoptera *Rhyacophila evoluta* (development constant 1150) lasts one year at 700 m altitude (Fig. 34). The eggs hatch in August-September. The autumnal and winter development of the aquatic phase is followed by the emergence of adults in July-August of the following year. At 1900 m, the cycle extends over 2 years, and at 2380 m it extends over 3 years. The cycle of Plecoptera *Isoperla viridinervis* (development constant 1780) extends over 2 years at 1850 m, and over 3 to 4 years beyond 2100 m.

Altitude	Duration of cycle	
700 m	1 year	01 2 3 4 (Winter) 5 N A
2000 m	2 years	0 (Winter 1) 1 2 3 4 (Winter 2) 5 N A
2380 m	3 years	0 (Winter 1) 1 2 3 (Winter 2) 4 5 (Winter 3) N A

Fig. 34. Biological cycle of *Rhyacophila evoluta* (Trichoptera) as a function of altitude. O, egg. 1 to 5, larval stages. N, nymph. A, adult (H. Decamps, 1967).

2.3. Conditions of life at altitudinal limits

In Margalef's formula (Chapter 4), under survival conditions of a population subjected to drift, the probability of drift increases when the duration of a biological cycle is long. At the upper limit of the altitudinal distribution, the number of individuals reaching the adult stage is very low or nil.

At 2750 m in the Central Pyrenees, it has been seen (Chapter 4) that losses caused by drift to populations of *Baetis alpinus* and *Rhitrogena loyolae* prevent an individual from completing its biological cycle locally, and that young larvae hatch from eggs laid by female adults that have come from downstream. At 2370 m, the survival of these two populations depends heavily on egg-laying females that are foreign to the environment, in a veritable process of reseeding. The population can be considered random.

A species with a biological cycle extending from 1 to 3 or 4 years as a function of altitude has a probability of drift multiplied by 3 or 4. The existence of shelter zones thus becomes critical for its survival. In the Estaragne stream, at between 1850 and 2380 m, migration in the sub-flow is high. It involves mostly larvae of Diptera Chironomidae (80% of individuals) or the young stages of other species. The majority colonize the first 10 cm of the sub-flow. The maximal larval density corresponds to high water from snow melt (Chapter 5).

2.4. Flight periods of insects that become flying adults

The flight period marks the completion of the cycle of insects that are aquatic at the larval stage and flying at the adult and reproductive phase. Three biological types appear, as a function of the flight period (Fig. 35):

- species flying in spring;
- species flying in autumn; and
- species with an extended flight period.

Fauna with spring flight, which is precocious, is characterized by a delay in emergence depending on altitude (Fig. 35A). This delay may reach 4 months in the Central Pyrenees, at between 500 and 2300 m, and emergence does not begin before July and beyond 2000 m. The flight period may be followed by a diapause of the egg, as with *Simulium timondavidi* (Fig. 33), or even a summer and winter larval phase in species with quiescence.

In reality, a good understanding of precocious flight periods would require that they be considered in relation to the rhythm of high and low waters rather than climatic seasons. Emergence in April occurs after the winter high water in a plains stream with a pluvial hydrological regime but before high water for a stream with a nival regime. In waters flowing from firn or glaciers, the emergence of certain species in June also occurs before the summer high water (glacier regime).

In species with autumnal flight, larval development occurs in spring and summer, and the egg or young larva stage in winter. Unlike precocious

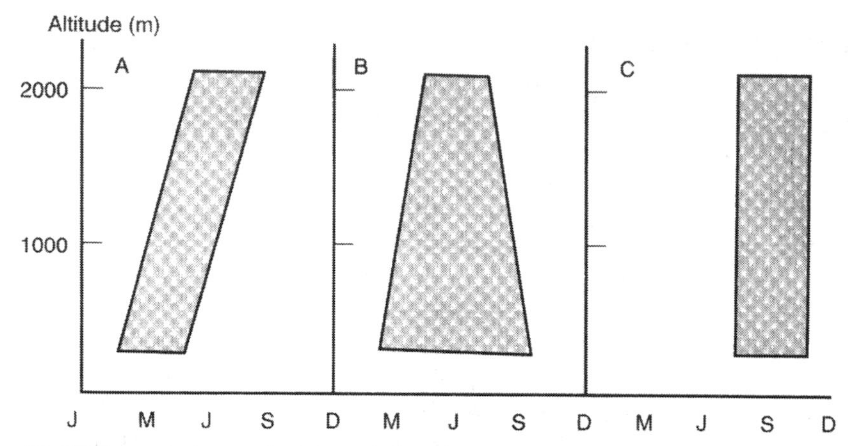

Fig. 35. Flight periods as a function of altitude in insects that become flying adults. A, species with spring flight. B, species with extended flight period. C, species with autumnal flight (H. Decamps, 1967).

flights, autumnal flight is independent of altitude or latitude (Fig. 35C). Factors other than temperature do intervene (perhaps seasonal factors such as photoperiods).

Species with staggered flight periods are numerous, especially at low altitude with polyvoltin cycles. On the Lissuraga stream, the flight period of adult *Baetis muticus* (which has three annual generations) extends over 10 months, from February to November (it is only 3 to 5 months in Britain, where there are fewer annual generations). *Ephemerella ignita*, a bivoltin species, flies 7 months of the year. Above 500 m, in the Central Pyrenees, the flight periods of adult *Hydropsyche pellucidula* and *Rhyacophila meridionalis* (polyvoltin Trichoptera) extend respectively from April to November and from February to December. The flight period becomes shorter with the altitude—June to August and August to September at 2000 m. The two species have by then become monovoltin.

Generally, for all insects that become flying adults, the flight period is shorter at high altitude. This results in the disappearance of polyvoltinism and the shortening of the warm season.

3. Conclusions: altitudinal distribution of fauna of running waters

For each species, the following must be taken into account:

— a temperature threshold for development;
— a tolerance zone;
— a resistance zone in which the duration of survival is increasingly limited;
— a cycle duration that depends on temperature and takes into account the development constant in the tolerance zone.

In addition, there is possible competition between similar species, which limits the actual distribution to less than the potential distribution, and there are species with diapause and others with quiescence.

For each species, an area of vertical distribution can be superposed that corresponds to a thermal area. It is difficult to compare overall the altitudinal distribution of species along a water course—each species having its own characters and limits. It is preferable to consider the succession of species that are phyletically similar—ecological vicariants, with the same needs in terms of substrate and food, but differentiated by their thermal preferences.

3.1. Altitudinal succession in Turbellaria

Crenobia alpina, Polycelis felina, and *Dugesia gonocephala* succeed each other from upstream to downstream (Fig. 36). The zone of cohabitation between two species is practically nil, and the replacement of one species by another results from competitive exclusion, as a function of the rate of increase and duration of survival.

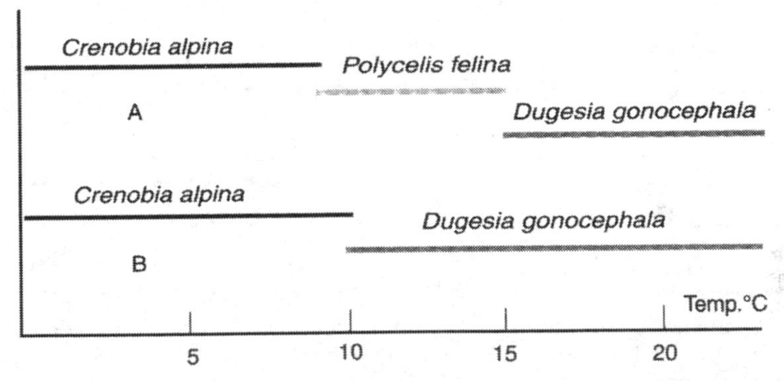

Fig. 36. Succession of Turbellaria on a longitudinal profile as a function of temperature: A, succession of three species. B, succession in the absence of *Polycelis felina* (E. Pattee, 1968).

Competitive exclusion is often cited by authors. It can be said to exist when species exploit the same food resources and when the absence of one species is accompanied by the extension of the distribution area of vicariant species. It will be seen (Chapter 11) that competitive exclusion seems limited to carnivorous species such as Turbellaria.

For example, *P. felina* is less resistant to being carried away by the current than the two other Turbellaria. In a rapid current, the altitudinal succession becomes *C. alpina–D. gonocephala*, the limit between the two species being at around 10°C. The same succession is observed in Corsica, where *P. felina* is not found.

In Romania, in the Banat massif, the single species of *Sericostoma* (Trichoptera) is *S. timidum*, which has a very wide area of distribution on the longitudinal profile of rivers. In Germany, *S. pedemontanum* colonizes the sources and the upper courses, while *S. timidum* is found in the low valleys.

3.2. Altitudinal succession in the Blephariceridae of the Central Pyrenees

The nine species of Blephariceridae (Diptera) of the Aure valley (Upper Pyrenees) have a diapause in the egg stage (Fig. 37). The altitudinal succession is characterized by large zones of cohabitation between species—unlike with Turbellaria.

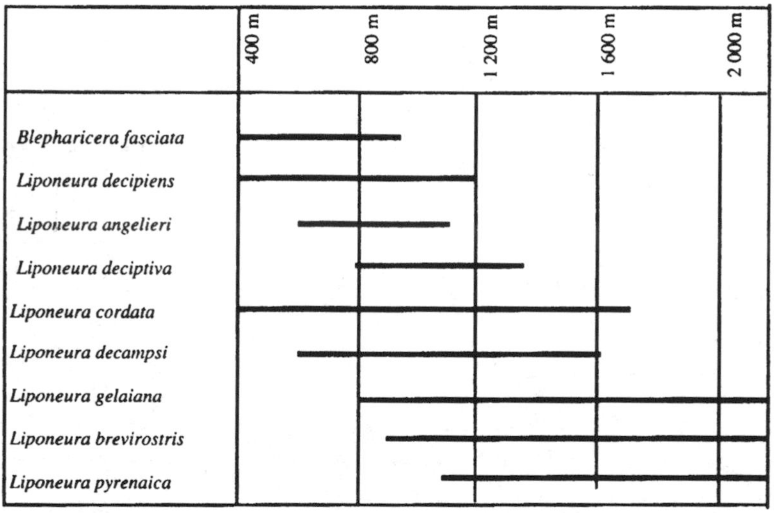

400 m
800 m
1 200 m
1 600 m
2 000 m
Blepharicera fasciata
Liponeura decipiens
Liponeura angelieri
Liponeura deciptiva
Liponeura cordata
Liponeura decampsi
Liponeura gelaiana
Liponeura brevirostris
Liponeura pyrenaica

Fig. 37. Altitudinal distribution of the Blephariceridae in the Aure valley (Upper Pyrenees) (J. Giudicelli and P. Lavandier, 1974)

Blepharicera fasciata colonizes streams up to an altitude of 750 m. Its distribution area coincides partly or totally with four species—notably *Liponeura decipiens* and *L. cordata* (which is found up to 1700 m). From 600 to 1200 m, *L. cordata* cohabits with three other species. At higher altitudes, two species cohabit—*L. brevirostris* and *L. pyrenaica* or even *L. gelaiana* and *L. brevirostris*.

All these species exploit the same food resources—the biological cover on the substrate—without resorting to competitive exclusion. Similarly, there are six species of *Hydropsyche* (Trichoptera) that cohabit on the upper Rhone upstream of Lyon and feed by means of a collector apparatus.

The succession in time of species that are similar and occupy a single territory allows them to exploit the same food resources while avoiding competition. Such resource sharing over time has been observed especially in certain Hydropsychidae in Britain and Canada. In Plecoptera

Nemouridae, optimal temperatures of embryo development and duration of incubation have been shown to result in a succession of similar species over a given time period.

3.3. Time-space successions in the Blephariceridae of Corsica

Seven species of Blephariceridae colonize the lower Tavignano and its tributary, the Restonica. They have phases of quiescence or diapause and are mono- or polyvoltin (Fig. 38). Their ecological characteristics result in faunal successions in space, along the longitudinal profile, and in time, over the course of the year.

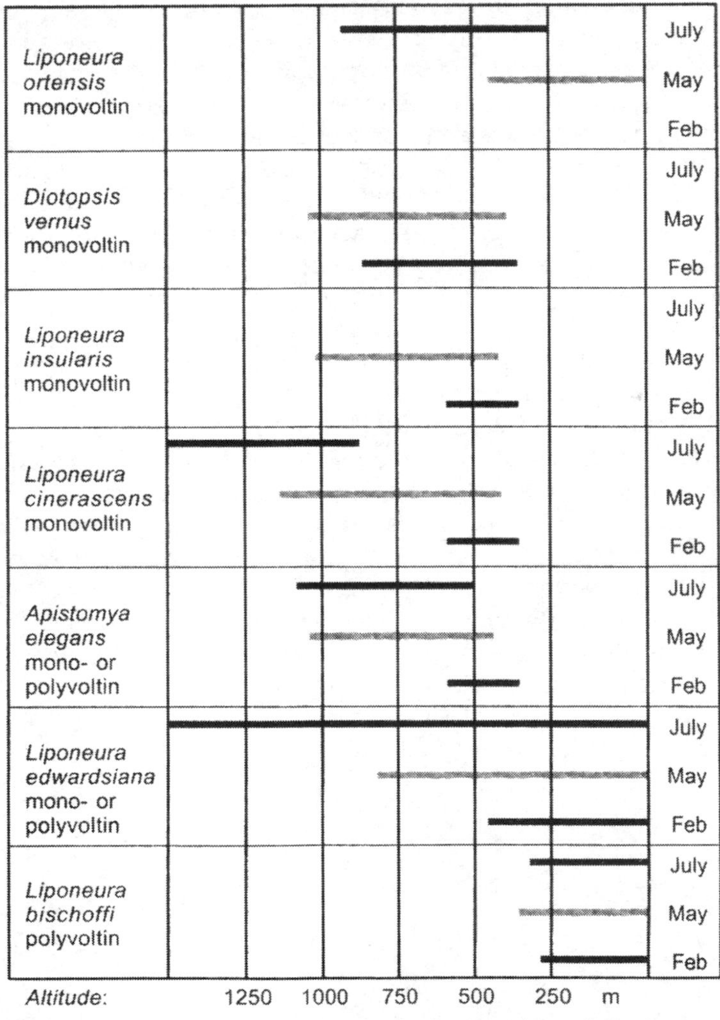

Fig. 38. Time-space distribution of the Blephariceridae of Tavignano, Corsica (G. Giudicelli, 1968)

Below 300 m, on the lower Tavignano, two polyvoltin species have continuous development: *Liponeura bishoffi* and *L. edwardsiana*. The generations overlap.

In February, at about 400 m altitude, there appear young larvae of three monovoltin species—*L. insularis, L. cinerascens,* and *Diotopsis vernus*—and one polyvoltin species, *Apostomya elegans*. In April, *L. cortensis* (monovoltin) also appears, while *L. edwardsiana* extends to the low valley up to 800 m. Six species, mono- or polyvoltin, constitute the spring fauna, which does not go beyond 1000 m altitude, and whose development is retarded as a function of the altitude.

The transition between spring and summer fauna takes place in June. The *L. cortensis* cycle is closed in the low valley and followed by a summer diapause. It is completed in July between 400 and 800 m. The cycles of *L. insularis* and *D. vernus* are also completed in July-August (winter diapause).

Above 400 m altitude, the summer fauna comprises three species. *Liponeura edwardsiana*, highly eurythermic, colonizes waters up to 1700 m—polyvoltin below 1000 m and monovoltine above 1000 m. Between 400 and 1000 m, it cohabits with *A. elegans*, also poly- or monovoltin as a function of the altitude. Beyond 1000 m, *L. edwardsiana* cohabits with *L. cinerascens*, a monovoltin species with a cycle that is retarded in relation to the altitude, which lives below 1000 m in the cold season and above that in the hot season.

Therefore, in Corsica there is a diversity of ecological types of Blephariceridae, unlike that observed in the Central Pyrenees. One species has a very localized distribution and continuous development—*L. bischoffi*. Two species have a wide altitudinal distribution, poly- or monovoltin according to the altitude—*L. edwardsiana* and *A. elegans*. Two monovoltin species, with diapause, have a cycle that occurs later in season as a function of the altitude. *Liponeura cortensis* does not go beyond 800 m, and *L. cinerascens* reaches 1700 m. Two monovoltin species appear with the spring fauna: *L. insularis* and *D. vernus*.

This faunal succession in time and space seems characteristic of streams of southern Europe, which have their source at high altitude but have high summer temperatures in their lower course. The diversity of ecological types is a result of the diversity of environmental conditions depending on altitude, season, and water flows.

10

Light, Salts and Dissolved Oxygen: Secondary Ecological Factors in Running Waters

While current and temperature are the two fundamental ecological factors in running waters, other factors may also play a role at variable degrees, namely, light, salts, and dissolved oxygen. Light and dissolved salts are important mainly for flora and dissolved oxygen for fauna.

1. Light and organisms in running waters

The ecological influence of light is different on fauna and flora. In plants, light is an energy factor that governs photosynthesis. In animals, it has only an indirect role, determining phases of activity, orientation, and seasonal rhythm (photoperiodism).

1.1. Light and aquatic plants

Part of the radiation that reaches the surface of the water is reflected. The other part penetrates the water, where it is progressively absorbed and dispersed, depending on the transparency of the water, until it becomes practically zero. The rates of extinction are not uniform. Infrared and ultraviolet rays are absorbed very rapidly while blue light penetrates deeper. Blue light is important in lakes, where light is involved in algal photosynthesis by its quantity (flow) and its quality (wavelength). Various pigments in algae are linked to various maxima of wavelength absorption, so that most of the visible spectrum is used, but selectively according to the class of algae.

Chlorophyceae present two peaks of absorption corresponding to chlorophyll pigments—one in the 400-440 nm band (indigo), the other in the 670-700 nm band (orange). In Cyanophyceae, phytoerythrine and erythrocruorine widen the bands of absorption, so that photosynthetic activity is possible over almost the entire visible spectrum.

Running waters are less deep than water in lakes. It is the quantity of light, rather than the quality of radiation, that governs photosynthesis. Optimal light intensity varies considerably from one species to another:

low (100 $\mu E/m^2/s$) in Cyanophyceae such as *Anabaena cylindrica*, high (more than 300 $\mu E/m^2/s$) in Chlorophyceae such as *Monoraphidium minutum* (Fig. 39). The result, in benthic algae, is a distribution of species according to the depth as a function of their degree of heliophily.

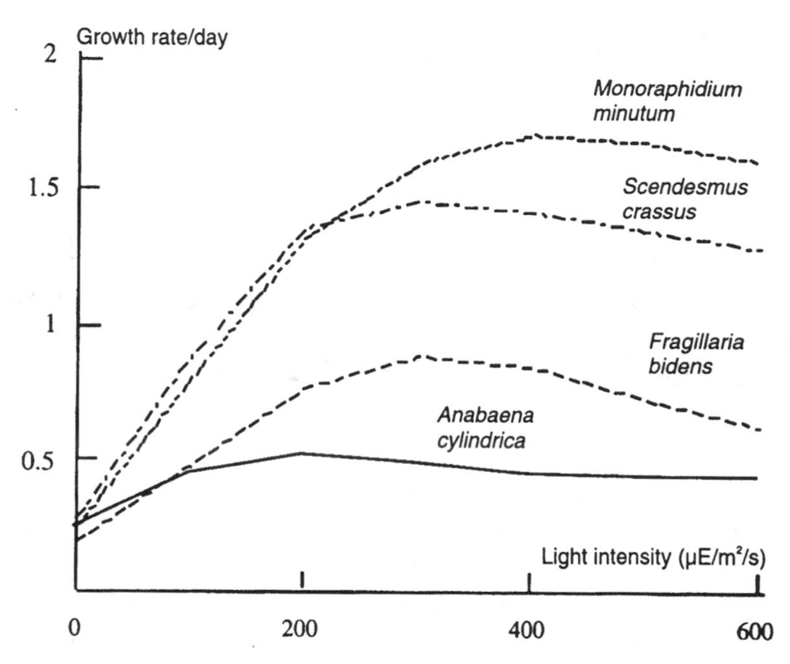

Fig. 39. Growth curves of four species of algae as a function of light intensity (A. Dauta, 1982)

As for phytoplankton, it develops in the *euphotic* zone, the depth of which corresponds to that of light penetration in the visible spectrum. The depth of the euphotic zone depends on both luminosity and water transparency. When the river is deep (potamal, canal reaches), an important factor controlling phytoplankton development is the ratio of *depth of euphotic zone* to *depth of the stream*. The deeper the stream, the longer an alga remains outside the euphotic zone and overall photosynthetic activity is reduced. Indeed, because of turbulence, a single cell successively occupies all the levels of the water column.

The distinction between heliophile and sciaphile species, of light and shade, is also found in macrophytes. Bryophytes tend to be sciaphile and colonize shaded streams or banks. Some species seem indifferent to light, such as *Fontinalis antipyretica* or *F. squamosa*. In reality, by living in deep areas of sunlit waters, they receive attenuated light.

Most entirely submerged Spermatophytes (helophytes) are heliophiles. However, even there, the depth affects the plant by limiting the available light. In the Dordogne river, downstream of the Argentat reservoir, *Ranunculus fluitans* colonizes the beds at a slope between 0.7 and 1.5%. It is a heliophile, and its maximum density corresponds to beds lower than 1.5 m. *Callitriche, Myriophyllum*, and Bryophytes develop in the shade of *Ranunculus*, on a stabilized substrate. At a depth of 2 m, *Ranunculus* and *Callitriche* disappear, and only a few plants of *Myriophyllum* and Bryophytes survive.

1.2. Light and fauna

In animals, light influences only behaviour. *Photoperiodism* (in animals and plants) is a seasonal rhythm determined by variations of the respective duration of day and night over the year. *Phototactism* refers to movements of organisms under the influence of light. It may be positive or negative.

In common trout, light determines orientation. Phototactism of fry is negative until the resorption of the vesicle, and then it becomes positive (free-swimming phase). Phototactism is again negative in old trout. It leads to a behaviour of searching for shelters and a daily feeding rhythm. At the Arctic circle, Salmonidae are diurnal during the winter and nocturnal during the summer.

Negative phototactism is most frequent in benthic invertebrates. If a shallow substrate is covered, most of the organisms come to the surface in a few minutes. Phototactism leads to a daily rhythm of activity, which is expressed by a daily rhythm of drift (Chapter 4).

In high-altitude streams, the drift, which is very slow, is arrhythmic while snow covers the bed. In summer, variations in light intensity (sunrise and sunset) play an essential role in the activity and peaks of drift. At sunrise, an overall reduction is observed in the density of the drift. The density then increases again for species with positive phototactism; it increases at sunset for species with negative phototactism. As in Salmonidae, phototactism may differ according to the age of individuals.

2. Dissolved salts

Dissolved salts or electrolytes, dissociated in the form of ions, reflect the geological nature of the terrain they cross (Chapter 1). Waters crossing acid terrain, granite, sandstone with siliceous cement, schist, gneiss, and basalt are poorly charged with electrolytes, particularly calcium. Conversely, waters crossing chalky terrain have high calcium and bicarbonate concentrations. Phosphates are rare in rocks and nitrous salts absent.

While current and temperature are factors that vary regularly from upstream to downstream, electrolytes, when waters cross only a single type of terrain, put a general stamp on the settlement, at least on the plants.

2.1. Electrolytes and aquatic flora

Flora, from algae to phanerogams, is more sensitive to electrolytes than fauna because it uses electrolytes for tissue development. As with terrestrial flora, aquatic flora is a better ecological indicator than fauna because, besides the current and temperature, it reflects the penetration of light and the qualities of the water and the substrate. Benthic algae and Bryophytes, in particular, are used as indicators of pollution.

Algae of acid or chalky waters are differentiated mainly by their capacity to absorb carbonic anhydride for photosynthesis—free CO_2 in acid waters and CO_2 of bicarbonates in chalky waters. Some macrophytes directly use the electrolytes of surrounding water (notably Bryophytes), others draw electrolytes of interstitial water by their root system.

Many plants are largely tolerant of calcium levels in water and constitute a floral base common to all water courses, acidic ones such as those of Brittany or the Vosges as well as the chalky streams of Causses or the Rhine plain. In addition to this common base are acidophile or calcicolous species, depending on the terrain crossed (Table 8).

No matter what the calcium concentrations are, plant groups also reflect concentrations of nutrient ions—nitrogen and phosphorus. These concentrations are indicators of the trophic level of the water, which may range from oligotrophic to eutrophic (Chapter 14), in terms of increasing

Table 8. Typology of Angiosperms of water courses of the Lozere from the source downstream, as a function of the nature of terrain crossed (J. Haury-CEMAGREF, 1992)

Upstream	——————————→		Downstream
Acidophiles, granites, gneiss, schists	*Chrysoplenium oppositofolium, Montia fontana, Callitriche hamulata, Caltha palustris, Juncus bulbosus, J. acutiflorus*	*Ranunculus aquatilis, R. penicillatus, R. peltatus, Myriophyllum alternifolium*	*Ranunculus penicillatus*
Common base	*Glyceria fluitans*	*Carex acuta, Nasturtium officinale, Phalaris arundinacea*	*Carex acuta, Epilobium hirsutum, Eupaterium cannabinus, Phalaris arundinacea*
Calcicolous, chalks, marl		*Ranunculus trichophyllus*	*Ranunculus calcareus*

concentrations of nutrient ions. Four groups thus appear in the acidic streams of the northern Vosges or of Brittany, and five in the chalky rivers of the Rhine plain. In the Vosges, the four groups succeed one another for nitrogen concentrations (in ammoniacal form) from less than 20 µg to more than 500 µg/l and phosphorus concentrations from less than 20 µg to 300 µg/l.

Calcium, nitrogen, and phosphorus determine not only plant distribution, but also productivity, since, along with light and carbonic anhydride, they control plant growth. However, they acquire true importance in a river only when the current is no longer a limiting factor (Chapters 8 and 14).

2.2. Electrolytes and aquatic fauna

Regulation of the internal environment in invertebrates gives them greater autonomy in terms of electrolyte concentration in the water. Salmonidae thus tolerate waters of pH 5 to pH 9, like the majority of aquatic species. These are the usual pH limits in natural waters.

Two cations may be significant by their excess or deficit: calcium and magnesium. Calcium reduces permeability of the cell wall and cuticle to water and ions, while magnesium has an antagonistic effect. Exchanges with surrounding water in crayfish are inhibited by calcium concentrations higher than 120 mg/l. However, these are extreme concentrations. Calcium is an important cation for plants that have an *exoskeleton* or a mineralized shell, the source of the calcium being the surrounding water. Rivers have no crayfish when their calcium concentration is less than 2.8 mg/l. The inflow of calcium (by the branchial walls) is less than the losses. The Amphipoda *Gammarus pulex* also disappears at below 5 mg/l of calcium, or even when there is an excess of magnesium in relation to calcium.

The testis of Mollusca *Ancylus fluviatilis* is transparent, nearly membranous, in acid water. The calcium reading (flow of uptake and losses) is negative below 1.04 µg/l of calcium, and even below 3 µg/l for populations acclimatized to very chalky waters. *Margaritifera margaritifera* is associated with waters that are deficient in minerals. However, the distribution of Molluscs can generally be said to be correlated to pH and to calcium hardness.

In invertebrates that have no exoskeleton or shell, calcium is not an important ecological factor. There are few documented cases of vicariance of species as a function of mineralization of waters.

3. Dissolved oxygen and fauna

Atmospheric gases are all soluble in water. Their solubility is a function of temperature, atmospheric pressure, and a coefficient appropriate to each gas. Gaseous exchanges occur at the water-atmosphere interface. Dissolved oxygen, together with carbonic anhydride, holds a particular interest, because its consumption (O_2) and its waste (CO_2) indicate the metabolism of organisms (Chapter 9). At a pressure of 760 mm Hg, the solubility of oxygen is 16.64 mg/l at 0°C, 11.26 mg/l at 10°C, 9.08 mg/l at 20°C, and 7.54 mg/l at 30°C.

Apart from dissolution from the atmosphere, oxygen is produced and carbonic anhydride is consumed in water during photosynthesis, and oxygen is consumed and carbonic anhydride is emitted by respiration. In shallow streams, in the rhithral, turbulence phenomena favour gas exchanges at the water-atmosphere interface. These exchanges are dominant in relation to those resulting from photosynthesis and respiration, so that oxygen concentrations in water correspond to the saturation as a function of temperature. The situation is different when the current is slowed down in the potamal and phytoplankton and plants can develop. Photosynthesis and respiration lead to a nycthemeral rhythm of oxygen and carbonic anhydride, from spring to autumn. There is oversaturation of oxygen during the day and undersaturation during the night. In the Lot river, summer development of phytoplankton induces oversaturation of 105 to 130% at 15 h, and undersaturation of 98 to 87% at daybreak. In dense vegetation, oversaturation is still higher, but at daybreak the concentrations may be close to anoxia. In this case, undersaturation prevails for a short time (a few hours of the night).

Oxygen consumption by poikilothermic organisms is a function of temperature. At a given temperature, it is also a function of the current.

Minimal concentrations needed by organisms are highly variable according to the species. Common trout colonize waters at minimal O_2 concentrations of 5 to 5.5 mg/l and a saturation rate of 80%. Experimentally, the lethal saturation rates for a number of fish species and benthic invertebrates have been measured (Table 9). The list of species, by decreasing order of saturation, corresponds simply to an upstream-downstream succession on a longitudinal profile, at currents that are increasingly slow and summer temperatures that are increasingly high. The lethal values measured are not habitually found in natural waters. This shows that the distribution of fish and benthic invertebrates in a stream is linked more to current and temperature than to oxygen concentrations. The problem of dissolved oxygen, on the other hand, occurs in polluted waters with a high organic matter content, when its consumption by aerobic bacteria is predominant in relation to exchanges at the water-atmosphere interface. It may occur temporarily in litter, the accumulation of dead, decomposing leaves, and sediments in which water circulation is reduced (Chapter 11).

In the hyporheic environment, dissolved oxygen concentrations no longer depend on exchanges at the water-atmosphere interface or on photosynthesis, but rather on the interference of surface and hyporheic waters and respiration of organisms. Several factors are involved in the

Table 9. Oxygen concentrations that are lethal for some species of fish and invertebrates (LC_{50}: lethal concentration in 48 h for 50% of individuals) (H. Ambuhl, 1959; U. Jacob et al., 1984; J. Arrignon, 1998)

	% saturation of oxgyen	Experimental conditions
Oncorhynchus mykiss (rainbow trout)	31.4	Survival to 84 h at 16°C
Cyprinus carpio (carp)	18.3	Survival to 84 h at 16°C
Perca fluviatilis (perch)	14.1	Survival to 84 h at 16°C
Tinca tinca (tench)	5.7	Survival to 84 h at 16°C
Baetis alpinus (Ephemeroptera)	82.5	LC_{50} at 15°C
Rhyacophila obliterata (Trichoptera)	70.2	LC_{50} at 15°C
Rhitrogena semicolorata (Ephemeroptera)	35	18°C, current 3 cm/s
Hydropsyche angustipennis (Trichoptera)	11	18°C, current 3 cm/s
Anabolia nervosa (Trichoptera)	9	18°C, current 3 cm/s
Ephemerella ignita (Ephemeroptera)	7	18°C, current 3 cm/s
Ephemera danica (Ephemeroptera)	0.56	LC_{50} at 15°C

oxygenation of hyporheic waters: speed of surface current, porosity, granulometry, and heterogeneity of sediment. In the Lachein stream in the Pyrenees, dissolved oxygen concentrations in hyporheic waters correspond overall to zones delimited by surface water flows (Chapter 5).

Stygobious species are as a rule less sensitive to low oxygen concentrations. However, only two species of Crustacea out of 22 colonize the reductive zone of the Lachein stream. As for the benthic species that colonize the hyporheic environment, they seldom penetrate very deep. They come close to the surface horizon of the substrate when low water and high summer temperatures lead to critical oxygen concentrations.

11

Food Webs and Energy Flows

This study has so far emphasized the difference between the functioning of terrestrial, oceanic, and lake ecosystems and that of running-water ecosystems, in which the transport of materials by the current leads to an upstream-downstream gradient of the biodegradation-biosynthesis cycle (Chapter 3). The primary origin of nutrients essential to the development of life is made up of organic matter from the watershed. Its biodegradation in water, analogous to that of a terrestrial soil, results in minerals needed for autotrophic plants (algae and macrophytes). Food sources for animals are therefore allochthonous as well as autochthonous (aquatic plants).

1. Allochthonous materials and their biodegradation

Allochthonous materials come from two sources: direct inputs as leaves from riverside plants fall into the water (in the autumn) and transfer to the water of litter that has fallen in the watershed. This organic matter is degraded by fungi and bacteria.

1.1. Inputs of allochthonous materials

Considered as a food source, inputs of allochthonous organic matter must be evaluated by unit area of the bed and by unit of time—m^2/year. Evaluated thus, the inputs decrease from upstream to downstream.

At the head of the watershed, in streams with a drainage order lower than 3, leaves from riverside vegetation fall on the entire width of the bed, which is not wide. The transfer of litter from the watershed by runoff water is a function of the slope. The total allochthonous input is high, about 250 to 650 g/m^2/year.

In the low valley, when the bed is wider, direct inputs fall only a few metres from the banks, and their quantity evaluated per m^2 of the bed is smaller. Upstream of Toulouse, France, the width of the Garonne is 160 m (order of drainage 7). The direct input of leaves affects only the first 15 m from each bank. The annual input is 42 g/m^2, while it exceeds 200 g at 2 m from the bank in just two months of autumn. However, this value is higher than that found generally in rivers with a drainage order of 6 or 7 (15 to 25 g/m^2/year in a river in Quebec, 23.2 g in the Thames, and 30 g in a river in Sweden).

In the natural state, the alluvial plain in the lower valley is covered with a forest in which the species vary as a function of the distance from the bank, elevation above the water table, and the flood plain (Chapter 12). Willows and alder are the trees found closest to the banks. Litter from the flooded part of the forest is partly carried away during high water in autumn and winter in streams with a pluvial regime, and in spring in streams with a nival regime. It is re-deposited during low water.

The relative quantities of direct inputs and inputs from the transfer of litter are variable. In the Garonne upstream of Toulouse, inputs of litter are estimated at 3.3 kg/year per metre of bank, of which 70% is direct input and 30% is lateral input from soil of the flooded part of the forest. However, in a Canadian stream, the lateral inputs represent only 10 to 17% of the total where the order of drainage is 5 and 6, and 40 to 50% where the order of drainage is 1.

The variability of inputs recorded by different authors is explained by the variability of the situations. Lateral input is fundamental in rivers such as the Amazon, flowing in the plains with a highly developed riverside forest and a large flood plain. However, in regions such as Western Europe, canalization, hydraulic plants, and flood-protection dykes have cut off rivers from their alluvial plain and reduced or eliminated the exchanges between the two environments. At present, for example, there remain only fragments of the old alluvial plain of the Garonne.

Dead leaves falling in the water and litter from the watershed are carried downstream by the current, then deposited in dead water areas, downstream of obstacles. They are again drawn into the current during high waters and deposited further downstream, then incorporated into the hyporheic environment and into the fine sediments of the potamal. It is during these phases of transport and redistribution that allochthonous organic matter is broken down into fine particles, degraded, and mineralized.

1.2. Biodegradation of allochthonous matter

There are two stages in the biodegradation of organic matter:

- A relatively short period of leaching. This stage leads to loss of soluble compounds, notably water-soluble polysaccharides and polyphenols.
- A slower phase of degradation by Hypomycetes and bacteria. This stage leads to final mineralization of organic compounds.

Fungi and bacteria act identically in terrestrial and aquatic litter. Hypomycetes play the most important role: they degrade cellulose and pectic polysaccharides of the cell wall and partly degrade lignin. This stage, which is manifested by a softening of dead leaves, is necessary to the colonization and consumption of litter by benthic invertebrates.

The action of bacteria only partly overlaps that of Hypomycetes. Few species degrade cellulose. Their essential role is rather the final degradation of compounds liberated by Hypomycetes and their oxidation.

The reduction of leaves into fine particles is the work of certain dilacerating invertebrates—Amphipoda Gammaridae, Isopoda Asellidae, some Plecoptera, and Trichoptera. The role of dilacerators—especially Gammaridae and Asellidae—is important in the Garonne upstream of Toulouse (50% of the biomass of invertebrates). It is much less so in a mountain stream, where Gammaridae and Asellidae are absent, while organic matter degrades more rapidly despite the lower temperature. The transformation of litter and dead leaves into fine particles undoubtedly results more from the abrasive action of the current than from the action of dilacerators, especially in streams with a steep slope.

The duration of the litter degradation cycle is determined by a combination of factors:

— plant species: willow leaves are degraded more rapidly than beech leaves, for example;
— temperature: degradation is faster in spring and summer than in winter;
— oxygenation of water, necessary to Hypomycetes and bacteria;
— abrasive action of the current;
— action of dilacerating invertebrates, which play the same role as the current in reducing litter and dead leaves to fine particles.

The relative importance of these different factors, depending on the sites and nature of plant sources, explains the variability in duration of the cycle of degradation of allochthonous organic inputs—2 to 10 months for alder, 4 to 21 months for poplar, 6 to 20 months for willow.

2. Autochthonous plant production

Autotrophic plants are those whose development in water is linked to solar energy. They are fixed (benthic algae and macrophytes) or in suspension (phytoplankton). In terrestrial, oceanic, and lacustrine environments, the ecological factors controlling plant production are light, temperature, nutrient salts, and dissolved trace elements. In running water, the current also has an influence, directly by eroding and limiting the extension of vegetation, and indirectly by determining the time of transit for algae in suspension (Chapter 7).

2.1. Phytobenthos and phytoplankton

The biological cover of the substrate, made up of benthic algae, also traps particulate organic matter. The benthic algae and particulate allochthonous matter plus the bacteria form a veritable micro-ecosystem characterized by rapid cycling of organic matter.

At the head of the watershed, in streams with a steep slope, erosion of the biological cover limits its development, as does the shade cast by the riverside vegetation. Plant biomass is small. From upstream to downstream, the current slows, summer temperatures are higher, as are the nutrient salt concentrations. Moreover, algal development is seasonal and occurs at low water levels: from spring to autumn in a pluvial oceanic hydrological regime, and from summer to spring in a nival regime. The winter biological cover of *Hydrurus foetidus* is characteristic of waters with a nival regime. Seasonal rhythm of the flow and temperature as well as slowing of the current from upstream to downstream results in a succession of the specific composition of phytobenthos in time and over the longitudinal profile of the streams (Diatoms, for example, are dominant in cold and fast-flowing waters). There is also an increase in algal biomass from upstream to downstream, which reflects the values observed by various authors.

The biomass of phytobenthos varies from 800 (streams of drainage order 1) to 2320 mg/organic carbon/m^2 (rivers of drainage order 7). Average productions are 5 mg/C/m^2/h (stream of order 1), 14 mg/C/m^2/h (order 3), 16 mg/C/m^2/h (order 5), and 29 mg/C/m^2/h (order 7). However, these are averages, which often mask enormous ranges between the extreme values: from 1 to 108 mg/C/m^2/h, for example, for a stream of drainage order 5 (average 16 mg).

Observation on a large scale clearly shows an overall increase in benthic algal biomass and its production on the profile along a stream. However, observation on a small scale shows a mosaic of microhabitats on the bed (Chapter 4). Each is a particular case, with its dominant factor controlling production—current, insolation, depth, turbidity. This is what is reflected by the ranges between the limit values.

The conditions under which phytoplankton develop are known (Chapter 7), and eutrophication is also seen in such conditions (Chapter 14). Overall, and in the same way as for phytobenthos, biomass and phytoplankton production increase from upstream to downstream. The biomass of a mass of water followed over 52 km covered in 10 days in August ranged from 0.4 to 1.6 mg/C/l. The increase appears regular. In reality, it is due essentially to a single nanoplanktonic Chlorophyceae, *Coelastrum reticulatum*, with a density multiplied by 100 in 10 days (500 to 50,000 cells/ml).

However, if, instead of following a flowing mass of water, regular samples from a single point are taken, a succession of microhabitats is found, equivalent in time to microhabitats of phytobenthos in space (Chapter 7).

This explains the variability of planktonic biomass and its production that has been observed by the authors. In the Seine and the Oise, for example, phytoplankton presents two annual growth peaks, one in April-May and the other in July (Seine) or September (Oise), with biomass of the order of 1.2 mg/C/l in the Seine and 2 mg in the Oise. In the Thames, four peaks exceed 0.8 mg/C/l, of which one reaches 7.8 mg. These data are difficult to compare or interpret given that a single mass of water cannot be followed, nor can its transit time and history be known.

2.2. Upstream-downstream gradient of detritic and algal particulate carbon

Theoretically, from upstream to downstream, there is a progressive movement from a heterotrophic ecosystem, analogous to soil, to an autotrophic ecosystem. The ratio of *primary productivity* to *respiration* (P/R, O_2 produced by photosynthesis to CO_2 emitted by respiration) is at first less than 1 (biodegradation being dominant), but it becomes greater than 1 when the primary production resulting from photosynthesis is preponderant.

The concept of *fluvial continuum* elaborated by R.L. Vannote et al. (1980) is based partly on this upstream-downstream gradient, with increased algal cover (and therefore increasingly fine organic matter). Such a gradient corresponds to an adjustment of consumers to the longitudinal evolution of food sources. This phenomenon can effectively be observed in summer, during low water. However, the concept of fluvial continuum is somewhat theoretical in that it does not take into account time and space discontinuities. The Rhine, for example, presents three successive longitudinal profiles—one at the source of Lake Constance, another from the source across the Rhine schistose massif, and a third that ends at the North Sea. The abrupt slope at mountain level (1800 m) leads to a disconti-nuity in many Pyrenean streams. A tributary loaded with suspended matter may introduce an autotrophic course into the heterotrophy, just as high water in the alluvial plain may recirculate organic matter from the flood plain.

Discontinuities of time are linked to the hydrological regime—the role of flow and temperature on algal development being of primary importance. Seasonal variations of algal biomass are generally greater than longitudinal variations at a given instant.

The hydrological regime of the Vire, in Normandy, is of the oceanic pluvial type—high water in winter (January-February), low water in summer. Algal carbon increases when the flow diminishes, while detritic carbon increases with the flow (Fig. 40). At a single point, the ecosystem is autotrophic or heterotrophic during the year, as a function of the flow.

Upstream of Toulouse, the hydrological regime of the Garonne is of the nival type: low water in winter and summer, high water in spring (May-June). Three peaks of algal growth are observed, corresponding to low water, and autotrophy ($P/R > 1$) appears only in August-September (Fig. 41). Overall, the P/R ratio is less than 1 over a year. Algal carbon represents only 27% of total particulate carbon, over a year. This percentage is nearly of the same order in the Oise (32%), but lower in the Vire (17% at km 40), the Thames (24%), or the Kennet (5.5%) in Britain. Finally, the concept of fluvial continuum considers the evolution of the P/R ratio over a regular longitudinal profile and during low water. It does not consider discontinuities in time and space nor does it reflect the hydrological cycle.

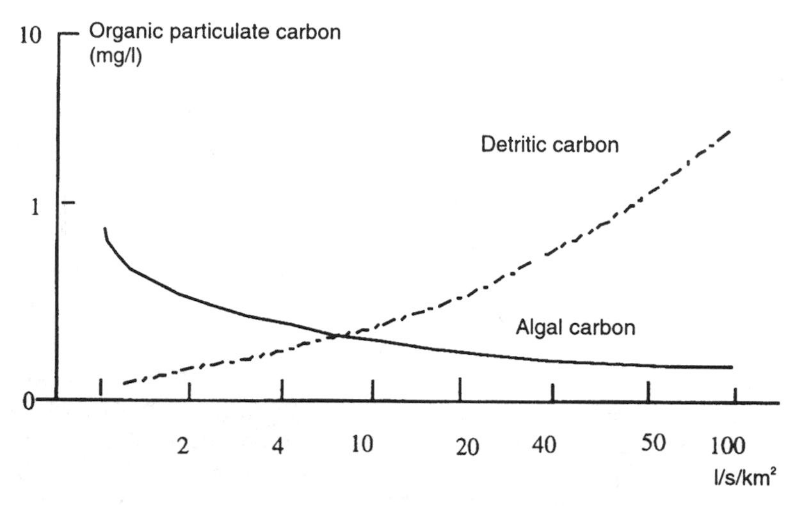

Fig. 40. Relationship between detritic carbon and algal carbon and the specific flow (l/s/km^2) on the Vire, at km 40 (S. Descy et al., 1984)

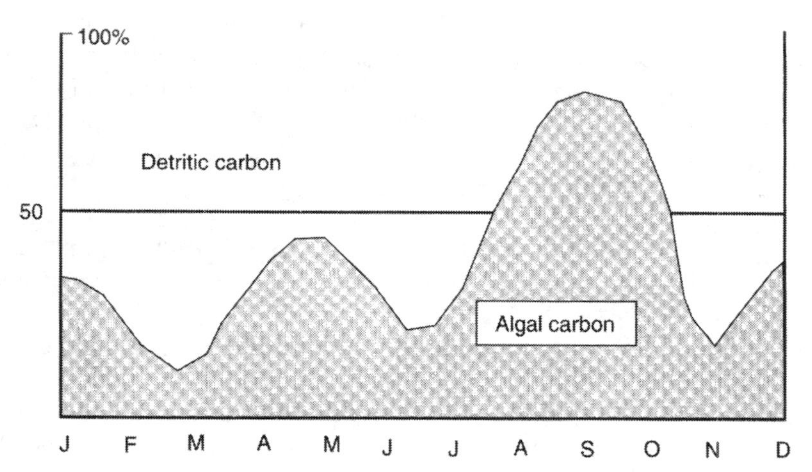

Fig. 41. Annual evolution of percentage of detritic and algal carbon in the Garonne, upstream of Toulouse (E. Chauvet, 1989)

3. Consumers

Allochthonous and autochthonous organic matter (the latter from algae, fungi, and bacteria) constitute the basis of the food webs of consumers, used either directly (by primary consumers) or at second hand (secondary consumers and carnivores).

3.1. Invertebrates

Depending on the species, invertebrates are primary or secondary consumers. The term *primary consumer* usually designates a species that feeds directly on plant production—detritivores feed on decomposing plant detritus.

In running waters, these two types of food are closely interlinked. The biological cover of algae traps coarse and particulate plant debris. This cover is in fact a combination of algae, organic matter, fungi, bacteria, and mineral particles. Invertebrates that scrape the substrate, such as Mollusca Ancylidae, Ephemeroptera Heptageniidae, or Trichoptera Thremmatidae, for example, ingest heterogeneous food and even small invertebrates together with the biological cover.

Only two taxa of the Estaragne stream (Upper Pyrenees) are strictly carnivorous—Turbellaria, which feed by sucking, and Plecoptera Perlidae, which have mouth parts of the grinding type. Fifteen other taxa have a more or less omnivorous diet. Trichoptera of the genus *Rhyacophila* are phytophagous at hatching and predators from stage II or III, but they consume the alga *Hydrurus* at the end of the season, as do Plecoptera Perlodidae, despite the presence of their habitual prey.

A carnivorous tendency is accentuated in the Trichoptera *Plectrocnemia, Allogamus, Drusus*, the Plecoptera *Arcynopteryx, Chloroperla*, and *Siphonoperla*, the Diptera Tanypodinae and Empididae. *Plectrocnemia* constructs a funnel-shaped net with which it rakes the contents. *Drusus* scrapes the substrate, ingesting the biological cover and the small invertebrates that colonize it. Invertebrates constitute 25% of the contents of the digestive tube of *Drusus rectus*.

Generally, in these omnivore-carnivores, diversification of diet is accompanied by increase in size. Similarly, the size of prey is a function of predator size. Omnivore-carnivores do not seem to have a choice of prey, apart from avoiding forms with a sheath (Trichoptera) or with highly chitinous teguments (Coleoptera). Euryphagy seems to be the rule. This allows the species to survive the abrupt disappearance of its prey (as during the massive emergence of insects). Nineteen taxa in the Estaragne streams are phytophagous and detritivorous. The Plecoptera Nemouridae and *Micrasema morosum* (Trichoptera) dilacerate degraded dead leaves and coarse woody debris. *Capnioneura brachyptera* (Plecoptera) grind plant debris and particulate detritus, while *Apatania stylata* grind detritus and browse the substrate (Diatoms are present in their digestive tubes).

Most of the other taxa of the river are microphagous browsers or scrapers of the substrate that ingest algae, particulate detritus, fungi, and bacteria—such as *Baetis* and *Rhitrogena* (Ephemeroptera), *Micrasema, Thremma*, and *Synagapetus* (Trichoptera), Orthocladiinae, Diamesinae, and Blephariceridae (Diptera).

The proportion of algae and detritus in the digestive tube depends on the microhabitat of each species and its feeding behaviour. Species living between pebbles consume more detritus than those living on the surface of the substrate, such as Blephariceridae. There is no *Hydrurus* in the microhabitat of *Thremma gallicum* and *Synagapetus insons*.

Some species collect their food by a filtering apparatus, for example Diptera Simuliidae and Tanytarsini, and Trichoptera *Philopotamus montanus*.

As for strictly detritivorous species, they live between pebbles or embed themselves in the hyporheic environment. There, organic detritus, along with fungi and bacteria, are the only food source. Leuctridae (Plecoptera) dilacerate plant debris, while Oligochaeta (except for Naididae) ingest fine particles. The presence of mineral particles in the digestive tube is also associated with a mode of life. Particles are notably abundant in species or in stages that penetrate the hyporheic environment, browsers or scrapers of the biological cover, and detritivores.

Finally, strict predators are few in number. Most of the invertebrates in running waters are omnivorous, with a carnivorous or herbivorous tendency. They are opportunists whose food varies with age, size, the season, and availability in their microhabitat. Under these conditions, the term *food chain* is not suitable for the designation of trophic relations in invertebrates. The term *food web*, or simply levels of consumers, would be more appropriate.

In the interstitial environment, the food web is simple, based on detritus and its decomposers, with a certain number of detritivorous consumers and some carnivores (Fig. 42). The peculiarities of the environment, in narrow interstices, hamper the action of predators.

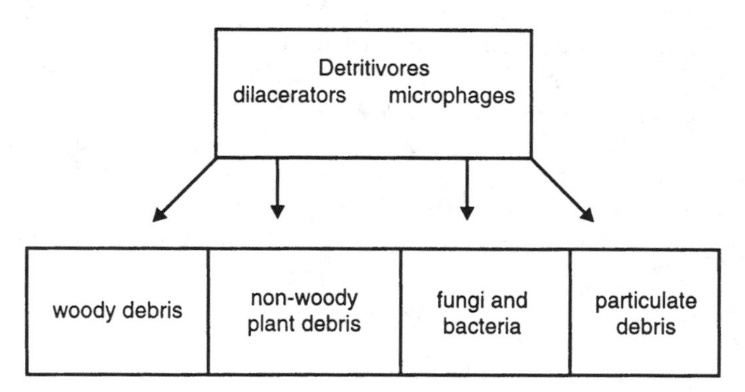

Fig. 42. Food web in interstitial environment

In the surface environment, the basis of the food webs is more diversified, and three levels of consumers can be distinguished (Fig. 43). The first is that of omnivores with a detritivorous or phytophagous tendency. The second are omnivore-carnivores, and the third strict carnivores. In streams with a low order of drainage, the dominant input of allochthonous materials favours detritivores and especially dilacerators. In the lower courses of rivers, with a high order of drainage, allochthonous inputs are greater near the banks and in dead branches—which are deposit

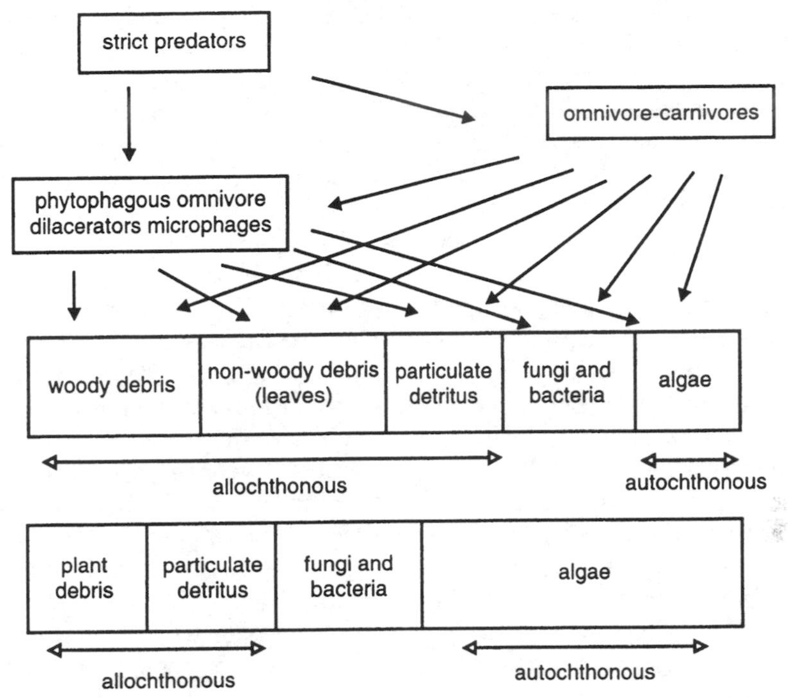

Fig. 43. Food webs of the superficial benthos: relative importance of food sources in streams of low order of drainage and banks of alluvial forests (A) and in rivers of the plains away from the banks (B).

zones. They are added to autochthonous inputs of macrophytes, mostly consumed after their senescence. The dilacerators are essentially Crustacea Gammaridae and Asellidae. Far from the banks, biological cover of algae is dominant in shallow waters and favours the phytophagous species, browsers and scrapers of the substrate—larvae of insects and Mollusca Gasteropoda. In the deep zones, where light penetration is limited, detritivores are again dominant.

The concept of fluvial continuum predicts, as a function of food resources, the high number of dilacerators in streams of a low order of drainage and of microphagous collectors in rivers of high order of drainage. If the entire course of a river is considered in a large-scale perspective, the image of an upstream-downstream continuum effectively appears. However, if the time and space discontinuities (high water, low water) are considered, the image of a continuum gives way to that of a mosaic of microhabitats with different food resources varying throughout the year. The input of alluvial forests and dead branches in the Rhone as well as the Garonne allows colonization of banks by dilacerators at the theoretical macrophage level. Facies in active current, in an open environment, favour browsing and scraping herbivores on the substrate, while detritivores colonize deep zones with less light.

The annual production of invertebrates can only be deduced from the numerical increase in population and weight increase of individuals. The ratio of production to biomass (P/B) expresses the renewal rate of biomass of a population. Most studies of production of invertebrates have been done on a single species, and the P/B ratio depends essentially on the duration of the biological cycle. It is high in polyvoltin species, low in monovoltin species, and still lower when the cycle extends over several years. Overall data are scarce and as variable as those of algal production. In the rivers of New Zealand, for example, the annual production of invertebrates varies from 0.8 to 72 g dry weight/m^2/year, depending on the sites and the available food—or a ratio close to 1 to 100. In Europe, researchers have observed levels of production between 3.3 and 110 g/m^2/year (this latter value was recorded from a river polluted by organic matter).

Annual production is low in high-altitude waters. It does not exceed 4.8 g/m^2/year (dry weight) at 2370 m in the Central Pyrenees, and 68% of this production comes from species with a biological cycle of 2 years or more. However, at 1850 m, the annual production is 7.6 g/m^2 and the semi-voltin species account for no more than 17% of that figure. In a small tributary of the Loire, at 800 m altitude, the production reaches 17 g/m^2. Mollusca Unionidae are practically the only invertebrates whose cycle extends over several years and represents 17% of this production.

In the plains, in the potamal, annual production is higher: 25.9 g/m^2 in the Thames and 57.4 g/m^2 in the Jihlava in the Czech Republic.

Overall, the annual production of invertebrates increases from upstream to downstream (just as for algae), in relation to the available food, but mostly to the shortening of the duration of biological cycles. These cycles correspond to more rapid growth and possibly to the succession of several generations in the course of a year. On the microhabitat scale, data from different publications are difficult to compare. At a very high altitude, for example, many species do not complete their biological cycle. In the plains, annual production is different close to the banks, in shallow zones with a biological cover of algae, and in deep zones where detritivores dominate (silt substrate).

Similarly, evaluation of annual production of different types of invertebrates gives results that are difficult to interpret. This is because there are no true food chains, but rather food webs with poorly defined contours, where the majority of species are omnivorous with a detritivorous or carnivorous tendency, and food regimes vary as a function of the stage of development, season, or a change in the microhabitat (Fig. 44).

3.2. Fish

Fish are located at the top of the food webs. They need protein-rich food, used for synthesis of body tissue and as a source of energy. As ininvertebrates, their basic metabolism is low because they are also poikilothermic.

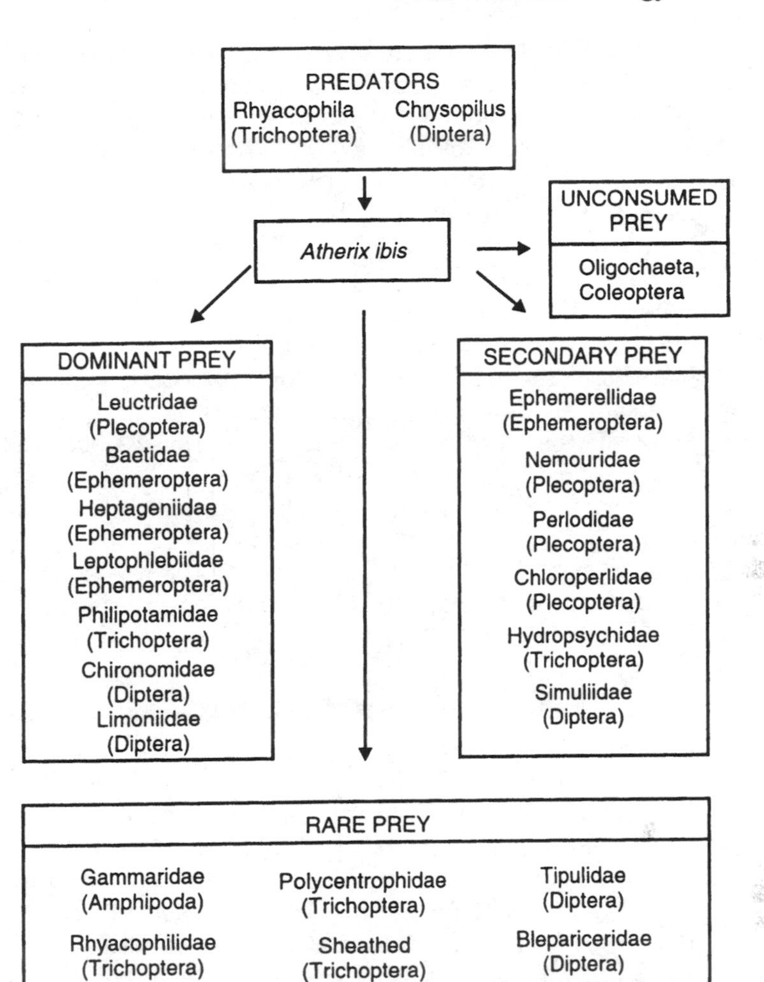

Fig. 44. Prey and predators of *Atherix ibis*, Diptera Rhagionidae (A. Thomas, 1981)

A certain number of species are omnivorous phytophages in tropical regions—the *Tilapia* in Africa, for example. In the Amazonian basin, out of 1800 known species of fish, more than 200 are granivores and frugivores. They feed for six months of the year in the flooded forest. At low water, they survive on the river bed, living on their reserves. On the Rio Negro, 80 out of 600 species feed on plant debris. In Asia, the chinese carp, *Ctenopharyngodon idellus*, is herbivorous. However, the majority of fish are omnivorous, even ichthyophagous, as are nearly all fish in the temperate regions.

Carp, tench, roach, bleak, and barbel are omnivore-carnivores (consuming worms, molluscs, crustaceans, and insects), without anatomical or physiological adaptations for feeding that are truly limiting. Young black bass exclusively consume invertebrates, while adults are also ichthyophagous (20 to 50% of invertebrates and tadpoles in their stomach contents). As for pike, although the fry is planktonophagous, it becomes ichthyophagous, just like perch or pikeperch, from the size of 4-5 cm.

The common trout is an omnivore-carnivore (consuming drifting and allochthonous benthic invertebrates). Allochthonous inputs may be very high during the summer. In a stream in the Upper Pyrenees, terrestrial invertebrates represent 57% of the prey in the stomach contents (70% in weight). Another possibility is consumption of small fish species (minnow) or juveniles. Chub, dace, and grayling have nearly the same diet. Mouth dimensions determine the maximum size of prey.

Finally, the same trophic levels are found in fish as in invertebrates. These prove, overall, the specificity and plasticity of these species in terms of preference for a type of food, but with sufficient eclecticism to draw from food that varies throughout the year. In running waters, fish, like invertebrates, are essentially opportunists.

The biomass of fish varies greatly according to the water course. For the common trout, the following values have been observed:

— 37 to 83 g/m^2 in Normandy streams (annual production 6 to 22 g/m^2);
— 2.7 to 34.4 g/m^2 in eastern France (1.7 to 12.9 g/m^2);
— 1.7 to 27 g/m^2 in the Central Massif (0.2 to 14.5 g/m^2);
— 1.8 to 32.4 g/m^2 in the Upper Pyrenees.

The question raised is whether or not these biomass values for trout reflect the availability of food. Their territorial behaviour and intraspecific and interspecific competition suggest that the biomass of a stretch of a river is linked to the quantity of available food. In similar sections of an English river, 12 to 280 trout fry were immersed per m^2. In all cases, in September, there remained only 8 troutlets per m^2, which results in a higher limit of biomass assumed per m^2. This leads to the concept of *reception capacity*, defined as the maximum density of a species or combination of species that an area of water can support during a given period. Since this reception capacity is evaluated in biomass, the number of fish per unit of area obviously decreases with their size. A trout of 25 cm occupies an area 100 times as large as does a troutlet of 5 cm.

However, the concept of reception capacity does not explain the ranges of biomass observed in French rivers. The biomass in place does not necessarily reflect annual production, which is the only relevant criterion (biomass can be modified by fishing or introduction of fish). Annual production depends on *biogenic capacity*, defined by L. Leger (1910, 1937) for Salmonidae as follows:

$$K \text{ (kg/ha/year of Salmonidae)} = 10 \text{ BC}$$

BC, biogenic capacity, is estimated from 1 to 10 as a function of abundance of benthic invertebrates. It does not take into account the total availability of food, in the form of drifting and allochthonous benthic invertebrates.

Primary benthic production, which can indicate the production of invertebrates, is correlated to temperature, illumination, and nutrient salts (expressed by conductivity and calcium or nitrate concentrations). These factors are often chosen for models explaining biomass of trout (reception capacity) or their annual production.

However, other factors, which are not inherently related to available food, are today considered in attempts to explain biomass and forecasts of reception capacity. For common trout, these are factors that describe their habitat (Fig. 45).

In a section of a river, the presence of a mosaic of facies, microhabitats—rapids, cascades, aprons, flats—and the free circulation of trout significantly influence the established biomass. The accessibility of microhabitats

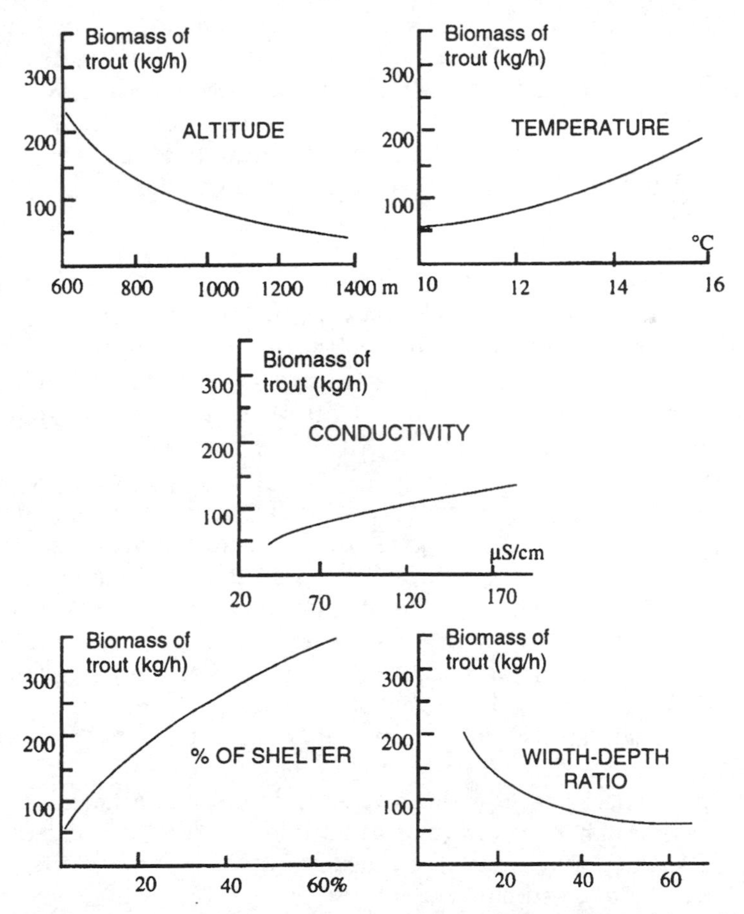

06) **Fig. 45.** Relationships between the biomass of trout (kg/h) and variables of habitat in streams of the Aure valley (Upper Pyrenees) (P. Baran et al., 1993)

propitious for reproduction (gravel substrates) determines the annual recruitment and thus the biomass. Different facies, which are fished separately, show highly variable biomass, while the flow of the drift is theoretically identical in contiguous facies. On a reach of a river, diversity of microhabitats seems to guarantee a high biomass, in providing places for hunting as well as resting places in a range of a few metres.

In rapidly flowing water, the notion that individuals save energy so as to remain in a steeply sloped area to capture prey seems to be dominant today. The width-depth ratio (Fig. 45) for a given flow corresponds to the slope. The percentage of shelter in relation to the total area indicates the value of the habitat. This value is of primary importance. In experimental conditions, the biomass of trout has been increased by introducing artificial shelters.

The reception capacity of rivers was estimated using Huet's slope-width index and calcium levels, but this does not take into account the value of the habitat. To do so, average current velocity, dimension of elements in the substrate, and possible shelters (including vegetation) all are integrated into predictive models on a vertical axis. This method, called the microhabitat method, aims to identify the needs of different fish species in terms of space, that is, of physical reception capacity. Some authors have also included temperature and nutrient salts with depth, current, and granulometry.

The microhabitat is the elementary cell, which is homogeneous for the parameters taken into account. The *useful weighted area* represents the sum of microhabitats of a sector, assumed from one coefficient (between 0 and 1) that expresses the value of the habitat for each fish species and each stage of a species. In common trout, for example, the value of the habitat is different for the fry, juveniles, and adults—apart from the spawning zones. In the streams of the Jura and the Alps, the useful weighted area corresponds to a biomass of about 58 g/m^2 for trout of more than 18 cm.

In basins of two Britanny rivers, the Scorff and the Elorn, the density of one-year juvenile trout (0+) is greatest in streams of drainage order 1 and 2 and that of two-year juveniles (1+) in streams of drainage order 2 and 3, while individuals three years and older massively colonize waters of drainage orders 3 to 20. This upstream-downstream distribution of different age classes results from the use of different microhabitats during the growth of common trout. However, at each stage, common trout are relatively sedentary and manifest territorial behaviour. Conversely, in the potamal, many fish species are extremely mobile.

This again poses the problem of the scale of perception of a water course—a mosaic of habitats or an upstream-downstream continuum. The addition of homogeneous units of mosaics, which result in the useful weighted area, allows estimates of the reception capacity of a segment of a stream and its piscicultural management. To estimate fish biomass on the overall scale of a water course, other methods taking into account periodic migrations or constant mobility are required.

A slope-width-calcium index and the microhabitat method have an applied objective: the piscicultural management of water courses. However, the merit of the two methods, fundamentally, is that they have shown that fish biomass and production must be considered more in terms of morphodynamic components of the environment than in nutritional terms.

4. Conclusions

The functioning of ecosystems is generally understood through the study of energy flow and its distribution in the food webs. It involves knowledge of the cycling time of nutrients, carbon, nitrogen, phosphorus, and other elements assimilated by organisms and entering into the composition of their tissues, subsequently de-assimilated and mineralized. Yet a running-water ecosystem is open and follows a longitudinal gradient, not a vertical gradient as in closed ecosystems (Chapter 3). The duration of the nutrient cycle—from organic to mineral form—corresponds to a distance covered by the water. This explains the *spiralling* concept of Elwood et al. (1981).

Estimations of biomass and its production are numerous. They express the quantity of energy present in the ecosystem and its rate of increase. Studies in the transfer of energy from one trophic level to another, and especially the overall balance of energy flows in a running-water ecosystem, are rarer and most often partial. There are many reasons for this.

In the first place, the metabolism of poikilotherms is low and does not require a regular and significant input of energy, unlike the metabolism of homeotherms. Moreover, it is dependent on the temperature, current, even oxygen levels and the substrate. The fauna of running waters is essentially opportunistic; the energy yield varies with the quality of the food ingested. All this makes the evaluation of energy transfers along the food webs difficult to quantify outside experimental conditions.

The study of fish species shows that food availability is only one of the elements from which biomass and its production can be estimated. Apart from food and temperature, the width-depth ratio of the bed and percentage of sheltered areas bring in morphodynamic factors. This is what the slope-width-calcium index and the microhabitat method indicate.

The principle of useful weighted area is valid not only for fish, but also for all benthic organisms. Current and temperature are the two primary limiting factors of the distribution of organisms, their conditions of existence, their biomass, and biomass production. Energy flow is a basic canvas for the development of communities rather than just a limiting factor of their biomass and its production.

As for plankton, its development is linked primarily to the transit time of water and the temperature. Nutrients and light become limiting factors of phytoplankton production when the transit time becomes very long and ceases in its turn to be limiting. In a development model for plankton in the Seine, the action of filtering rotifers on the density of phytoplankton appears only during low water, and more so in the summer. The prey-predator relationship, which corresponds to an energy transfer, is episodic. The phytoplanktonic biomass, moreover, with a few exceptions, is only a trophic dead end that is practically unusable by consumers (contrary to what is observed in stagnant waters).

In terrestrial ecosystems, which are closed, plant and animal successions in time, from the pioneer phase of colonization to the climactic phase, correspond to the development of regulation strategies and optimal use and distribution of energy inputs. Over the course of successions, energy and available matter are substituted progressively for climatic and physical factors as the first limiting factors of biomass and its production.

Running-water ecosystems, which are open, present no successions in time, but only successions in space, from upstream to downstream. They remain in the pioneer stage of colonizers, where physical factors are largely dominant and where each community depends on the one upstream. Each community uses and transforms nutrients transported by the current. The strategy of organisms is not one of optimal use of energy and matter carried and redistributed by the current, but an adjustment to physical and morphodynamic factors and recolonization (e.g., dispersal by flying adults in insects, high rates of reproduction). The rapid repopulation of environments destroyed by a flood or an exceptional pollutant indicates how easily areas that are accidentally destroyed are recolonized.

Despite incessant transformations during the year, running-water communities prove to be relatively stable from one year to another. This results more from the stability of morphodynamic conditions and the regularity of the hydrological regime than from the energy flow and its use. These are the characters of pioneer ecosystems and the causes of difficulties in establishing balances of energy and matter flows.

In the concept of fluvial continuum, water courses are described not only by matter flows, but also by their morphodynamic variables and temperature, which present a continuous longitudinal gradient. Three classes of water course are defined by their order of drainage: 1 to 3 (upper basin), 4 to 6, and large streams of an order higher than 6. Flora and fauna respond to the gradient by adjustments and a succession of communities.

This concept has a major effect on the biotic adjustments of communities along the longitudinal gradient, while biotic factors do not appear dominant. Invertebrates are most often opportunistic. However, if the adjustment to morphodynamic conditions of the water course is ranked first, the concept remains useful in understanding and comparing the longitudinal successions of communities. As such, it is linked to Illies and Botosaneanu's zonation of water courses (Chapter 13), and, beyond the mosaic of microhabitats, indicates a more holistic perception of a longitudinal profile of a water course.

12

From Upstream to Downstream: Ecological Zonation of Water Courses

Since the strategy of organisms living in running water is mostly an adjustment to morphodynamic factors and temperature, and not a regulation of populations by the flow of energy and available nutrients, no time successions are observed from pioneer ecosystems to climactic ecosystems. There are only longitudinal successions of pioneer species from upstream to downstream. These are based on the relative position of each species to all the others (Table 10).

Table 10. Average position, from upstream to downstream, of Coleoptera *Hydraena* and Elminthidae in the Garonne basin (C. Berthelemy, 1966)

1. *Hydraena emarginata*	12. *Limnius opacus*
2. *Esolus angustatus*	13. *Dupophilus brevis*
3. *Limnius perrisi*	14. *Esolus parallelepipedus*
4. *Riolus violaceus*	15. *Elmis maugetii*
5. *Hydraena saga*	16. *Hydraena gracilis*
6. *Hydraena truncatula*	17. *Riolus cupreus*
7. *Elmis aenea*	18. *Oulimnius tuberculatus*
8. *Hydraena pygmaea*	19. *Oulimnius troglodytes*
9. *Hydraena minutissima*	20. *Esolus pygmaeus*
10. *Limnius volkmari*	21. *Limnius intermedius*
11. *Elmis coiffaiti*	22. *Stenelmis caniculata*

Corresponding to successions of species are successions of communities, i.e., sets of species reacting similarly to a set of factors that constitute the experimental environment. When no factor is truly limiting, the species adjust more to the overall complex than to a particular factor of it. The *temperature-current velocity* pair occupies, as has been seen, first place in the hierarchy of factors. The same species may occupy habitats in rapid current when the temperature is high and in slow current when the temperature is low, for example, the common trout.

Species even migrate into the hyporheic environment in rapid current and at low temperature and return to the superficial horizon of the substrate in slow current and at high temperature.

In the hydrographic basin of the Doubs river, six communities can be identified from upstream to downstream through a factorial analysis of correspondences (Fig. 46). This statistical succession of ecological types indicates the organization of certain regularities in the settlement of running waters.

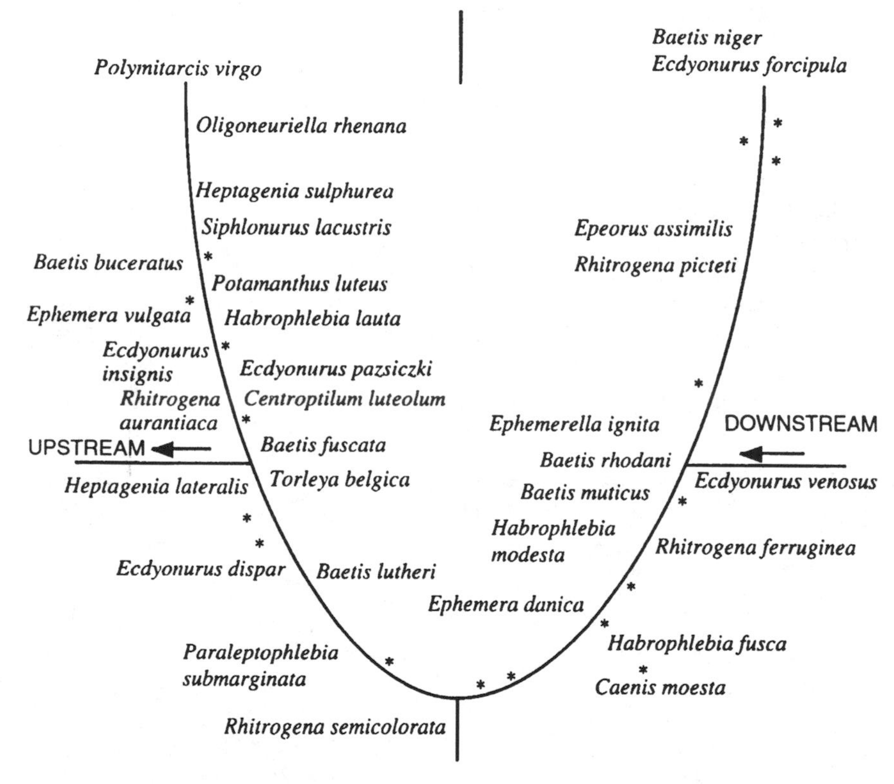

Fig. 46. Upstream-downstream succession of Ephemeroptera in the Loue stream (Jura) according to a factorial analysis of correspondences (J. Verneaux, 1973)

Many authors have attempted to classify running waters using either an upstream-downstream zonal typology or a typology of microhabitats. The two typologies are two modes of perception of a water course: an upstream-downstream gradient or a mosaic of microhabitats. These two modes are not mutually exclusive.

1. Types of microhabitats

Apart from the sources (see section 2.1), the different microhabitats of running waters depend on morphodynamic parameters, slope, current, depth of bed, and granulometry of the substrate (Table 11). They can be classified as shown in Fig. 47.

Table 11. Classification of sediments as a function of their diameter

Rocks	> 500 mm
Rubble	200 to 500 mm
Stones	20 to 200 mm
Gravel	2 to 20 mm
Coarse sand	0.2 to 2 mm
Fine sand	0.02 to 0.2 mm
Powder	0.001 to 0.02 mm
Precolloids	0.1 to 1 μm

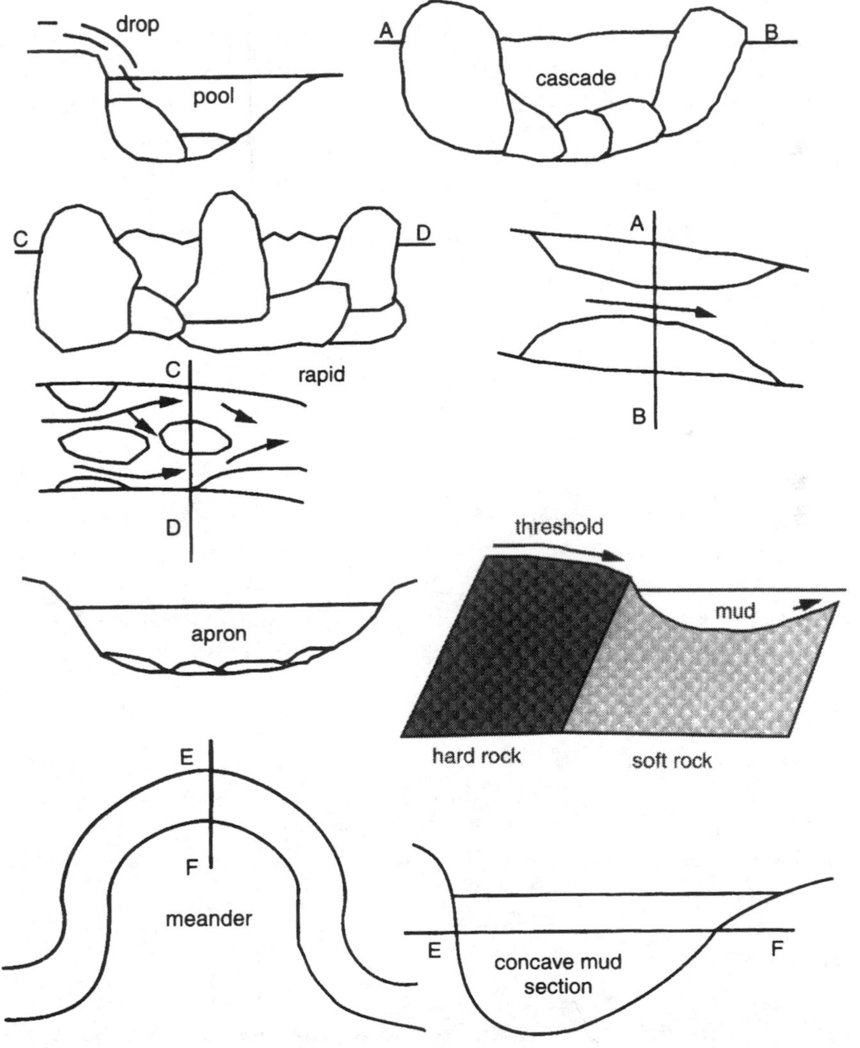

Fig. 47. Different types of microhabitats in running water (J.R. Malavoi, 1986)

1.1. Falls and cascades

A waterfall is created when the mother rock rises to the surface and a geological accident leads to a crack and a high leap. A cascade is created when the bed is encumbered with rocks and stones, the width of the submerged section is reduced, and current velocity is accelerated. At the foot of a fall or cascade, the force of the current erodes the bed. Thus, a pool is formed, in which the energy stored by the water is dissipated. The current slows down and the granulometry of the substrate is fine, consisting of gravel and coarse or fine sand.

1.2. Rapids

Over steep slopes (> 4%), the rapid and turbulent current creates swirls. The granulometry of the substrate is coarse, consisting of stones and rubble that can be rolled along during high water. This leads to random distribution of the materials of the bed and chronic instability of the substrate. Behind the stones, counter-currents may deposit finer materials, i.e., gravel and sand. The depth of the water is variable—shallow when the mud section of the bed is wide, more than 40 cm deep when that section is narrow.

1.3. Aprons

In a widened bed, the water spreads out and its velocity (< 40 cm/s) as well as its turbulence decreases. Aprons are zones in which materials are deposited.

1.4. Flats

In a wide bed with a low slope, the current becomes uniform (< 40 cm/s), as does the depth (< 40 cm). Turbulence is practically nil, and the largest materials in the substrate are pebbles and gravel carried during high water. At low water, flats are zones in which fine material is transported, rather than deposit or erosion zones.

1.5. Muds

Muds are deep zones (> 60 cm) with a very slow current during low water. They develop during high water, and the granulometry of the substrate is variable. Muds may originate from deepening of concavities in the meanders or near obstacles in the subsidiary bed (thresholds of mills, locks) on reaches with a low slope.

1.6. Channels

In straight or slightly curving reaches, channels are characterized by a U-shaped profile and great depth. Depending on the velocity of the current at low water, a channel is called lentic (slow current) or lotic.

1.7. Lones

Over reaches of meanders, the curves may form a closed loop and result in dead branches. These branches ultimately become totally isolated from the main bed and their settlement evolves towards a lake type.

2. Upstream-downstream zonation

Two upstream-downstream typologies have contributed significantly to progress made in the ecology of running waters in Europe: that of M. Huet (1949) and that of J. Illies and L. Botosaneanu (1963).

Huet's typology refers to certain morphodynamic factors (slope-width) and to fish, which are highly mobile and have a wide ecological tolerance. Illies and Botosaneanu's zonation is based on morphodynamic factors, temperature, and benthic invertebrates, which are seen as better biological indicators than fish. It indicates discontinuities in the upstream-downstream settlement during rapid modifications in morphodynamic factors (flow, width and depth of the bed). These discontinuities are found at the confluence of streams or rivers of equal size and of the same order of drainage (Fig. 48). Three zones, themselves subdivided, succeed one another from the source

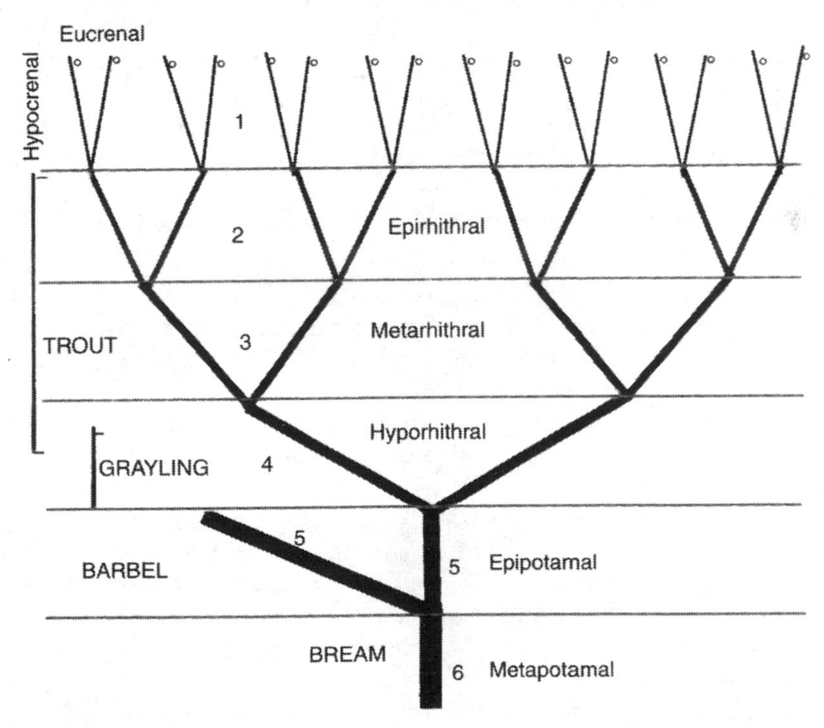

Fig. 48. Illies and Botosaneanu's and Huet's zonations and order of drainage on a watershed (Illies and Botosaneanu, 1963)

to the estuary of a water course—crenal, rhithral, and potamal—and their communities constitute the *crenon*, *rhithron*, and *potamon*.

2.1. Crenal

Strictly speaking, the term *crenal* designates the sources themselves (eucrenal) as well as the rivulets of drainage order 1 that flow from them (hypocrenal). Depending on the origin of the water and its flow pattern, four types of sources can be recognized:

— Helocrene sources are simple depressions of the terrain that interrupt the underground layer. The flow is slow and the depression is in fact a pond or a bog.

— Limnocrene sources are also depressions in the soil forming a basin. However, the water flows under pressure and the flow may be rapid. The substrate of helocrene or limnocrene sources, consisting of fine sediments, is similar to that of lakes. It is colonized by macrophytes such as *Caltha palustris, Montia rivulorum, Stellaria uliginosa*, mosses, and Sphagnum. The fauna is also of a lacustrine type (Oligochaeta, Mollusca, Crustaceae, Diptera Chironomidae larvae).

— Rheocrene sources gush over a slope, following a fracture in the continuity of the subterranean layer. Mountain springs are rheocrene sources. The substrate is coarse, consisting of rocks, rubble, pebbles, and gravel. Rubble and rocks are covered with Bryophytes. The biological cover is poorly developed, and the fauna is related to that of the rhithral, consisting of larvae of insects that become flying adults dominating a petricolous environment, to which are added Copepoda, Ostracoda, and Oligochaeta in the Bryophytes.

— Resurgences are sudden reappearances of underground waters, which may involve an entire water course.

Sources and resurgences are located at the limit between underground and surface domains, and the crenon may arise partly from ground water, particularly Crustacea Amphipoda (*Niphargus*), Isopoda, Copepoda, and Ostracoda.

The single point common to the different types of sources lies in the narrow annual thermal range, which is expressed by seasonal cycles and a poorly defined eucrenon.

The temperature is a function of altitude and latitude: from 7.8 to 9.8°C below 500 m, 4.5 to 8°C at 1400 m, in the Black Forest, and –4°C at most in sources in the subalpine and alpine stages of the Alps and the Pyrenees.

Annual thermal amplitudes are a little higher in the hypocrenal. It is as heterogeneous as the eucrenal and is in fact a mosaic of microhabitats analogous to those of the rhithral. However, it involves streams of order 1, in a narrow bed with low flow. The microhabitats are highly fragmented, occupy a restricted space, and are well-defined. On a steep slope, the hypocrenon resembles the eucrenon and rhithron, with Bryophyte flora in stable and shaded substrates. When the slope is low, the communities are close to those of helocrene and limnocrene sources.

The madicolous habitat is a particular case of the crenal. The source is a simple oozing of the mother rock or a big rock, and the water depth does not exceed 2 mm. The temperature, which is relatively stable at the site of the oozing, subsequently presents wide daily fluctuations depending on the insolation. Certain madicolous species breathe the air directly by means of stigmata located in the part of the body exposed to the atmosphere (larvae of Diptera Psychodidae and Stratiomyidae).

The species diversity of crenobiont organisms is low. In the Black Forest, out of 26 species of Hydracarids of rheocrene sources of the mountain stage (850-1100 m), a single one is cited in 11 sources out of 13, and 14 are cited in a single source. At the subalpine stage (1300 m), a single species colonizes 24 sources out of 25, and 9 colonize fewer than 3 sources. This indicates that the settlement of the crenal is largely random, especially at high altitude, because of the small area of microhabitats and the isolation of sources, which limits faunal exchanges. Each crenon has more linkages with the rhithron that follows it than with other crenons.

2.2. Rhithral

The term *rhithral* designates streams and tributaries found in mountains or hills with slopes greater than 0.15% and with a rapid and turbulent current. It begins with streams of drainage order 2, issuing from confluents of rivulets of the hypocrenal. The rhithral conforms to Huet's trout zone. With turbulence, it is characterized by an annual thermal average not exceeding 20°C in summer (in temperate regions) and a coarse substrate consisting of rocks, rubble, pebbles, and gravel. The dissolved oxygen concentration is equal to saturation (gaseous exchanges with the atmosphere are facilitated by turbulence).

The coarse substrate favours circulation in the hyporheic environment, as well as penetration into this environment by benthic fauna. When the hyporheic environment is in contact with ground waters, it is also colonized by stygobious fauna (Chapter 5).

Organisms of the rhithral (Table 12) are for the most part stenotherms, with clear adaptations to life in the current—in terms of body shape, organs of fixation (suckers, hooks), adherence to the substrate (mucus), or behavioural adaptations (such as burrowing). Rapid current and turbulence cause, apart from erosion of the substrate, a significant drift of organisms. This favours larvae of insects with a phase of airborne dispersal, allowing constant recolonization of the environment. Invertebrates lacking a flying stage (Turbellaria, Oligochaeta, Mollusca, Crustaceae) are confined to habitats sheltered from the current (Bryophytes, hyporheic environment, dispersal pools) or stick to the substrate (e.g., Mollusca Ancylidae).

Table 12. Families and sub-families of insects and Hydracarids that preferentially colonize the rhithral

Ephemeroptera	Baetidae, Heptageniidae
Plecoptera	Nemouridae, Leuctridae, Capniidae, Perlidae, Perlodidae
Coleoptera	Hydraenidae, Elmidae, Helodidae
Trichoptera	Rhyacophilidae, Glossosomatidae, Philopotamidae, Polycentropidae, Brachycentridae, Limnephilidae
Diptera	Blephariceridae, Simuliidae, Diamesinae, Orthocladiinae, Corynoneuridae
Hydracarids	Thyasinae, Protziinae, Sperchoninae, Torrenticolinae, Axopsinae, Aturinae

The most rheophilous and the most stenothermic species have a relatively limited geographical distribution, probably for the same reasons as species of the crenal, i.e., isolation of high-altitude watersheds. Even for the flying adults of insects, low atmospheric temperatures do not favour movement over long distances, especially the crossing of mountain chains. In the deep valleys, adults can disperse only in parallel with the water course.

In the rhithral, three successive zones are recognized—the epirhithral, the metarhithral (Huet's trout zone), and the hyporhithral (grayling zone). They are based on hydrodynamic characters and the settlement. From one zone to another, the settlement may differ qualitatively or statistically according to the taxa or more or less sudden environmental variations.

Overall, species diversity of the rhithron increases from upstream to downstream. In the Neste d'Aure valley (Upper Pyrenees), Fisher's diversity index (S species = $\alpha \log_e$ (number of individuals/α)) varies from 1 to 1.8 in the epirhithral to 4 in the hyporhithral for petricolous Hydracarids. This is expressed in the increased number of species when the slope and current diminish and the number of degree-days increases.

The epirhithral corresponds to torrents of drainage order 2, over the steeper slopes of the watershed, where the bed is narrow and shallow and the flow reduced. This bed is a mosaic of small, well-defined microhabitats dominated by falls, cascades, and rapids (on slopes greater than 4%) and a granulometry consisting of rocks, rubble, and pebbles. Fine sediments are deposited only in dispersal pools and behind rocks (counter-currents).

The only macrophytes of the epirhithral are Bryophytes, fixed on the mother rock or on stabilized rocks. On narrow beds, the shade of riverside trees favours Bryophyte development, but not that of the algal cover (which is also limited by the corrosive action of the current). *Hydrurus foetidus* and Diatoms are dominant algae at low water.

In the torrent of the Estaragne stream (Upper Pyrenees), 168 out of 195 invertebrate species have an airborne dispersal stage. Ephemeroptera, Plecoptera, Trichoptera, and Diptera Chironomidae, representing 61% of species, contain 90% of individuals of the rhithron. The substrate is chronically unstable in the rhithral. Bryophytes and the hyporheic environment are the only relatively perennial habitats. Of the hyporheic fauna of the Estaragne stream, 90% consist of Diptera Chironomidae and young larvae of Ephemeroptera and Plecoptera. The constant erosion of the substrate leads to the thinning of alluvia above the mother rock and the absence of connections with ground waters. The hyporheic environment is reduced to just benthic fauna.

For invertebrates, the essential source of food is organic matter from the watershed (leaves, litter) and the decomposers of that matter (fungi and bacteria).

The metarhithral (drainage order 3) results from the confluence of two or more streams of the epirhithral. It is still a zone of the torrential type, but the bed is wider and the flow at least double that of the epirhithral. Microhabitats of the fall or cascade type regress in favour of rapids and aprons. The aprons are spawning zones for trout.

The zone that is shaded by riverside trees affects only part of the bed, and the penetration of light in low water favours development of algal cover, which in turn is a food source for invertebrates that browse and scrape the substrate. The metarhithral remains a zone of dominant erosion. However, the thickness of the alluvia may lead to a veritable water table and the presence of stygobious fauna in the hyporheic environment.

Depending on the taxa considered, the metarhithral appears to be a distinct, or even an intermediate, zone situated between epi- and hyporhithral. Thus, for the Diptera Orthochladiinae *Eukiefferella* and *Cricotopus* of the Lot river (Table 13), only four species out of 17 characterize the crenal and the epirhithral, two the metarhithral, and two the hyporhithral. The presence of other species can be represented by two Gauss curves that overlap in the metarhithral. This also seems to be a transition zone for Diptera Simuliidae of the Neste d'Aure (Upper Pyrenees) and the Fulda river (Germany). On the other hand, in the hydrographic network of the Carpathians, groups of Trichoptera are as clearly individualized in the metarhithral as in the epi- and hyporhithral.

The hyporhithral is a grayling zone—mountain or piedmont streams (drainage order 4) resulting from the confluence of at least two torrents of the metarhithral. The annual temperature ranges are 15 to 18°C. Microhabitats of the torrential type are in regression, while aprons and flats dominate, and hollowing out (in meanders and around obstacles) results in muds. The area of these formations increases. There are successions of aprons and flats, with substrates of pebbles, gravel, and sand. The hyporhithral, a zone in which these materials are deposited, is

Table 13. Distribution of genera *Eukiefferella* and *Cricotopus* (Diptera Orthocladiinae) on the three levels of the rhithral and the crenal of the Lot (H. Laville, 1981)

		Crenal	Epirhithral	Metarhithral	Hyporhithral
Eukiefferella	*dittmari*	+	+		
	brevicalcar	+ +	+		
	fuldensis	+ +	+		
	claripennis	+ +	+ + +		
	discoloripes		+ +	+	+ +
	coerulescens	+ +	+	+	+
	lobifera	+	+ + +	+ +	+
	calvescens	+ +	+ +	+ + +	+
	clypeata	+	+ +	+ + +	+
	devonica	+	+	+ +	+
	similis	+ +	+	+	
	pseudomontana			+	
	ilklyensis			+	+ +
	pottasti				+
Cricotopus	*pulchripes*	+	+		
	annulator	+	+ +	+ +	+ + +
	tremulus		+ +	+ +	+ + +
	curtus		+	+	+ +
	trifasciata		+	+	+
	sylvestris		+	+	+ +
	similis			+	+ +
	triannulatus		+ +	+ +	
	vierrensis			+	+ + +
	bicinctus			+ +	+ +
	albiforceps				+
	flavocincus				+

characterized by the size of the hyporheic environment and its relationship with the ground waters.

Insolation and the substrate of pebbles and gravel limit the growth of Bryophytes in developed zones with a stabilized and shaded substrate (e.g., diversion canals, bridge piles). Conversely, insolation and slow current favour heliophile phanerogams, which appear along the banks and aprons.

In the Scorff river in Brittany, France, the *Oenanthe crocata* group characterizes the river bed from about 10 cm and in a current of about 35 cm/s. *Callitriche hamulata* and *Ranunculus penicillatus* colonize flats rather than aprons (depth of about 25 cm and current of 30 cm/s). The algal cover is well developed at low water.

With respect to fauna, among species characteristic of the hyporhithral, the importance of taxa with a purely aquatic life must be noted —Oligochaeta, Mollusca, and Crustaceae (Copepoda, Ostracoda, Gammaridae, Asellidae).

In an environment with less turbulence, recolonization by air no longer appears to be an essential factor in the permanence of the rhithron.

2.3. Potamal

A result of the confluence of mountain or hill rivers flowing in an alluvial plain with a gentle slope (< 0.15%), the potamal accords with Huet's barbel and bream zones. While pebbles, gravel, and sand are deposited in the hyporhithral, silt is deposited in the potamal.

Microhabitats are of the channel or mud type (at meanders) and are relatively deep, with a non-turbulent flow. They are larger than microhabitats of the rhithral, stretching over long reaches with a substrate of pebble covered with silt at low water (lotic channels) or silt only (lentic channels).

The difference between potamal and rhithral lies in a modification in the hierarchy of hydrodynamic factors. The current remains the organizer of the substrate, but it is no longer a primary limiting factor. The temperature may exceed 20-21°C in summer. In the plains streams, which are shorter and flow from cold sources, Salmonidae remain (Huet's slope-width rule is not valid for Normandy streams, for example). The high summer temperatures favour polyvoltin species. A seasonal succession is also observed, of stenothermal species of cold water sources from autumn to the beginning of spring, and warm water species in summer (Chapter 9).

The potamon is made up of three different ecosystems: plants, plankton, and benthos.

The fauna, whether it lives in the benthos or the plants, is related to that of stagnant waters, which indicates a modification in the hierarchy of ecological factors: the transport of materials and organisms is reduced at low water.

In the potamal of the Trieux stream (Brittany) appear *Potamogeton, Elodea canadensis, Myrophyllum alterniflorum, Nuphar lutea*, and *Sparganium emersum* (Chapter 6). These are characteristic plants of flats and channels, submerged or with floating leaves, and their distribution is correlated to depth and light penetration (Chapter 10). The algal cover develops close to the banks, at shallow depths, and in the vegetation.

The algae suspended in the waters of the rhithral are practically all of benthic drifting form. Along with the potamal appears a true plankton, the development of which is primarily linked to the water transit time (Chapter 8).

The extent of photosynthesis (algae and plants) is expressed by high oxygen production during the day, while respiration and oxidation of organic matter constantly consume oxygen. The absence of turbulence and the depth of the water limit gas exchanges at the water-atmosphere interface. The result, at low water, is a daily cycle of dissolved oxygen, oversaturation during the day and undersaturation at the end of the night.

The fauna of the potamal is similar to that of stagnant waters in the high incidence of taxa without an airborne dispersal phase (Table 14): these include Oligochaeta, Hirudinae, Mollusca, Crustacea (Copepoda, Ostracoda, Cladocera, Gammaridae, Asellidae, Athydae), and Insecta (Heteroptera, Coleoptera Dytiscidae, Haliplidae, etc.). The taxa with an airborne dispersal phase are different from those of the rhithral: Plecoptera are rare, and burrowing species or those living in plants dominate among Ephemeroptera, Trichoptera, and Diptera.

Table 14. Principal taxa that colonize essentially the potamal

Turbellaria	*Dugesia tigrina, Dendrocoelum lacteum*
Oligochaeta	Tubificidae, Lumbriculidae, Lumbricidae, Aelosomatidae
Achaeta	Piscicolidae, Glossosophonidae, Hirudinae, Erpobdellidae
Mollusca, Gasteropoda, Bivalves	Valvatidae, Viviparidae, Hygrobiidae, Planorbidae, Physidae, Lymnaeidae, Dresseniidae, Sphaeridae, Unionidae, Margaritiferidae
Crustaceae	Copepoda, Ostracoda, Cladocera, Gammaridae, Asellidae, Athyidae
Hydracarids	*Pilolebertia, Hydrodroma, Limnesia, Hygrobates fluviatilis, Forelia cetrata, Wettina, Piona, Mideopsis*
Ephemeroptera	Ephemeridae, Polymitarcidae, Potamanthidae, Caenidae, Leptophlebiidae, *Ephemerella*
Odonata	Zygoptera, Anisoptera
Heteroptera	Geocorisae, Hydrocorisae
Coleoptera	Dytiscidae, Hygrobiidae, Hydroptilidae, Haliplidae, Helephoridae
Trichoptera	Hydroptilidae, Ecnomidae, Phryganeidae, Leptoceridae
Diptera	Chironomidae, Culicidae, Ceratopogonidae, Tabanidae, Stratyomiidae, Limoniidae

Two types of settlements are clearly differentiated: that on the bottom and that in the plants. Bottoms of silt or pebbles covered with silt are colonized by a fauna that scrapes the substratum, such as Mollusca Gasteropoda, filtering organisms such as Mollusca Bivalva, or limivores such as Oligochaeta. Many species are burrowers, including Oligochaeta, Ephemeroptera *Ephemerella* and *Polymitarcis*, Mollusca Sphaeridae, Unionidae, Margaritiferidae, Diptera Tipulidae, Limoniidae, Ceratopogonidae, and Tabanidae.

The interstices of the hyporheic environment are sealed by silt. The hyporheic fauna, which moves between the interstices, is replaced by a burrowing fauna, which moves by displacing the sediments.

The fauna in the vegetation is similar to that found in lake vegetation. The vegetation is a source of food and provides shelter for fish (Chapter 6).

In the potamal, three successive zones can be distinguished: the epipotamal, corresponding to Huet's barbel zone, the metapotamal or bream zone, and the hypopotamal, corresponding to the zone of influence of the sea, with a brackish-water fauna.

Unlike stenothermic invertebrates of the crenal and the rhithral, eurythermic species of the potamal have a wide geographical distribution. This is probably due to largely open valleys and to the larger ecological adaptability of species, which allows them to colonize different environments. Among the Trichoptera, for example, groups of species of the potamal have a wider ecological adaptability than those of the rhithral.

2.4. Illies and Botosaneanu's zonation and the concept of fluvial continuum

Any attempt at classification involves subjectivity, and overlapping can always be observed between neighbouring zones or microhabitats. In ecology, all research begins with a typology, whether of identification of species, communities, or ecological factors. The typology provides a reference that allows comparison between different water courses and comprehension between researchers. It is also indispensable for administrators.

Illies and Botosaneanu's zonation, the best known in Europe, is based on morphodynamic factors as well as on groups of invertebrates. The morphodynamic factors are those of Huet's zonation, i.e., slope-width, complemented by flow, depth, order of drainage, and temperature. Statzner concluded from a study of running waters that hydraulics is the essential factor and temperature a secondary factor in ecological zonation. In Europe, this conclusion corresponds fundamentally to Illies and Botosaneanu's zonation.

The use of invertebrates allows for a more precise zonation than the use of fish alone, since fish are mobile and have a wide ecological adaptation. The two zonations partly overlap: epi- and metarhithral correspond to the trout zone, hyporhithral to the grayling zone, epipotamal to the barbel zone, and metapotamal to the bream zone. The classification of microhabitats fits into a longitudinal zonation. Obviously, microhabitats of the fall, cascade, or dispersal pool type dominate in the epirhithral, aprons in the hyporhithral, and channels in the potamal. The appearance of the microhabitat mosaic changes from upstream to downstream.

Illies and Botosaneanu looked for *nodas*, which are discontinuities in the morphodynamic factors due to abrupt changes in flow and bed width at the confluents of streams of similar size. These nodas are observed particularly in canopy massifs or plateaux of Hercynian Europe. However, in the glacial valleys of the Central Pyrenees, the abrupt 500 m drop at the mountain stage is also a noda, separating epi- and metarhithral. At the subalpine stage, the typical zonation can be erroneous. For example, the Estaragne stream (Upper Pyrenees), above 1850 m, flowing over steep slopes in a small valley open to the north, corresponds to an epirhithral. Conversely, at the same altitude, but in a sunlit valley facing south, the Estibère stream crosses several lakes and peat bogs. It is colonized by a middle-altitude fauna that finds suitable temperature conditions there—especially at points where lakes empty into the river. The thermal regime thus has a heavy influence on the location of communities. An inversion can also be observed in the succession of zones. A stream flowing on a high-altitude plateau is composed of a succession of aprons, flats, and muds, followed by an epirhithral. This does not undermine the concept of longitudinal zonation, the zones being identifiable even when their succession is not regular.

It has been shown (in Chapter 11) that, in the concept of fluvial continuum, water courses are described by their morphodynamic variables, temperature, and the biotic adjustments of communities. Yet, the adaptation strategy of running-water organisms is expressed in their mode of life rather than their food regime. In the hierarchy of ecological factors, current and temperature occupy the two primary places, while food quality appears secondary. If, in the concept of fluvial continuum, morphodynamic factors are given an essential place, this is very close to Illies and Botosaneanu's zonation. A water course appears to be an upstream-downstream continuum. If there are nodas or discontinuities, the communities are distinctly isolated. In the absence of discontinuities, the groups of species are of statistical value. However, in either case, the Illies and Botosaneanu zones can be identified.

From the overall picture of a water course, the opposition between rhithral and potamal can be perceived. At a smaller scale there appear subdivisions of the rhithral and the potamal, as well as the crenal. At a still smaller scale, a point on the longitudinal profile can be defined by its experimental environment complex: altitude of the source, altitude of the experimental stations, surface area of the watershed (or even bed width, distance between source and experimental environment, order of drainage), and hydrological regime (or summer temperature).

Finally, on a tiny scale, the mosaic of microhabitats can be perceived. This shows a local instability, in contrast to the overall stability observed in the large ecological units of the upstream-downstream succession.

3. The alluvial plain and its zonation

The banks of a water course follow an upstream-downstream longitudinal gradient, just like the water course itself. On the upper watersheds, in an erosion zone, the banks are most often sloped and the bed deep. During floods, the bed does not spread out much, and the flow is ensured by just the rise in water level and the increased speed of the current. Materials are transferred from upstream to downstream.

As the slope diminishes, the bed spreads out over the less raised banks. The erosion-deposit cycle of materials is no longer merely longitudinal but also lateral, from each side of the channel, during high water. Thus, a periodically flooded alluvial plain is constituted in which the bed widens, together with the secondary channels and dead branches. In the valley, raised terraces reflect hydrological regimes and former levels of the river. Of the seven terraces of the Rhine, downstream of Strasbourg, six were formed after the Neolithic period.

3.1. The alluvial plain

The granulometry of materials follows a lateral gradient, in keeping with the soil level and distance from the bank. The coarsest materials—pebbles, gravel—are deposited close to the channel, where the current is the most rapid during high water. Silt is deposited at the highest levels, at the boundary of the flooded zone, during a slower current. As the waters recede, silt is also deposited close to the channel, but it is recirculated during the following rise in water level. The horizontal zonation of materials, in the alluvial plain, is analogous to their longitudinal zonation in the river, from the rhithral to the potamal.

3.2. Vegetation on the banks

The bank vegetation follows an upstream-downstream gradient, as do the nature of banks, the substratum, and organisms living in the stream. On the upper watershed, the soil is higher than the average level of the water table, and the source of supply to plants is essentially rain water. The bank forest is independent of the stream—consisting of oak, beech, spruce, or pine depending on the climate. Conversely, it is of critical significance for the ecology of the stream itself by the shade it casts and the matter—twigs, leaves, organic soil detritus—washed down by runoff, stored, or transported by the current. As the slope of the banks diminishes, and even more so in the alluvial plain, the soil level approaches the ground waters and the banks become periodically flooded.

Three factors determine the distribution of plants on the alluvial plain: the periodicity of floods, the nature of the alluvia, and the depth of the water table.

On gravel banks of tress beds and on channel banks, dried up during low water and very close to the ground water, amphibious plant groups develop, including *Phalaris arundinacea, Melilotus alba, Mentha aquatica,* and *Alisma plantago-aquatica*. These are pioneer species, on substrates that are reshaped with each flood.

On the upper French Rhone, alluvia of sand and silt covering the gravel present five plant groups that succeed one another according to the average depth of the water table (Fig. 49). As the water table falls lower, its importance in supplying water to plants diminishes, in favour of rain water. After the amphibious group of *Melilotus alba*, willows are the closest to the banks and the ground water on soils with gravel and sand, the silt being swept away with each flood. Alder then grows in soils with sand and silt (*Alnus incans*) or in clay and peat soil (*Alnus glutinosa*). The oak and ash group is located 2-3 m above the ground water, in silty soil. Elm appears in former channels filled in with clayey silt, and poplar in soils poor in organic matter.

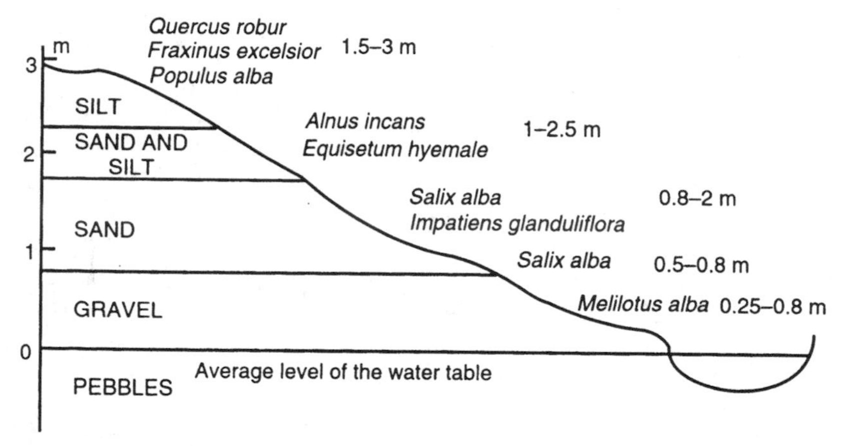

Fig. 49. Plant successions in the alluvial plain of the upper French Rhone in accordance with average depth of the water table (G. Pautou, 1984)

Channel banks that are unstable and regularly flooded over long periods (where *Melilotus* and willow are found) appear to be pioneer ecosystems, with low species richness. Where floods are more episodic and brief, species richness increases (to more than 50 woody species in oak and elm groups of the upper Rhone).

The raising of the alluvial plain from the banks is not regular, nor is the water flow. High water may hollow out the alluvia at certain points and accumulate materials at others, just like the river itself. Depressions remain in the sites of former channels. The distribution of plant groups is rarely as regular as that described, and it corresponds most often to a mosaic similar to that observed in the crenal and the rhithral.

The alluvial forest can be tens of kilometres wide around tropical rivers flowing in the plains, with a short dry season, such as the Amazon. At present, in temperate, densely populated countries, the alluvial forest is reduced and fragmented. In Alsace, for example, the alluvial forest of the Rhine covered 20,000 ha in 1840; in 1930 it covered no more than 16,000 ha, and at present it covers only 7000 ha (of which more than half is reforested). The alluvial forest of the Garonne, downstream of Toulouse, remains only in the major bed flooded 30 days a year at flows of 650-750 m^3/s. The zones flooded decennially (1500-2700 m^3/s) or exceptionally (up to 4500 m^3/s) have become farm land.

3.3. Transfers between channels and the alluvial plain

The same kinetic energy that brings in alluvial deposits can also erode them and transfer them to the channel. Thus, exchanges take place from the river to its alluvial plain and from the plain to the river.

The water of the lower valley is most often loaded with mineral salts, nitrates, ammonia, and especially phosphates. By filtering through the flooded alluvia, it carries essential nutrients to plants. Silt, clay, and detritus deposited at the edge of the flood plain at high water will remain on the entire alluvial plain as the waters recede. Organic matter, which emerges during low water, is mineralized more rapidly than in the channel; it helps enrich the soil in mineral salts and contributes to the growth of alluvial vegetation. This vegetation absorbs mineral salts and stores them in the form of wood for periods of months or years. In this way, it also helps purify the surface and ground waters, notably by the elimination of nitrates and phosphates, at least partly. On the Garonne, the denitrification capacity of the alluvial forest has been evaluated at 50 mg nitrogen/m^2/day. Similarly, the nutrients carried in flooded dead branches later contribute to the development of vegetation.

In autumn, leaves and dead wood falling from trees are carried by the next flood along with clay and silt and are transferred to the channel, just as on the slopes of the upper watershed. The ecosystem, autotrophic in the middle course, can thus become heterotrophic again, at least temporarily, in the alluvial plain by means of the organic matter load and the reduction of photosynthesis due to turbidity.

Channel-plain transfers also involve fauna. The alluvial plain corresponds generally in the channel to barbel and bream zones, in which many species are poorly rheophile. Secondary branches and dead branches are sites for shelter and reproduction during high water. Rheophile species, such as barbel, chub, roach, and dace, reproduce in the secondary branches of tress beds, while species that are hardly rheophile use secondary branches that are blocked upstream and dead branches (e.g., bream, tench, catfish, rudd, pike). The invertebrates of the channel also migrate towards the flooded zone.

These faunal transfers from the channel to the alluvial plain corres-pond to transfers from the plain to the channel—such as terrestrial insects and benthic and planktonic invertebrates of dead branches. They carry surplus food to fish in the channel. In the upper French Rhone, the drift of the Isopod *Asellus aquaticus* towards the channel is multiplied by nearly

400 when the flow increases from 400 to 630 m^3/s (annual average flow 450 m^3/s).

3.4. The mobile littoral concept

The alluvial plain appears to be a bed complementary to the channel, a reserve of inertia that restrains the violence of floods and a supplier of nutrients to the river.

Flash floods, by definition of short duration, are accidents in the functioning of the ecosystem endured by organisms. The same is not true of regular, seasonal, or predictable floods of long duration. These promote the development of strategies by which organisms of the channel effectively exploit the flooded zone.

In tropical and subtropical regions, the dry season is short, between December and March in the Orinoco, for example. The fish of the flood plain (170 species) generally have an annual reproductive phase synchronized with the beginning of high water. Their nourishment, as well as that of juveniles, is ensured by forest products (fruits, seeds, leaves) and flood lakes (plankton and benthos). Many species receive nourishment only on the flood plain, subsisting on their reserves during low water (Chapter 11, section 3.2).

Studies on the river Danube over a period of nearly 40 years show a correlation between piscicultural production and the duration of flooding: 500 t/year for 20 days of flooding, 750 t for 100 days, and more than 1500 t for 200 days. This increase in piscicultural production accords with an extension of the useful weighted area during a long period. The alluvial plain is no longer just an interface, an ecotone or transition zone between the terrestrial and aquatic environment. It is integrated with the river, the littoral of which has become mobile. This mobile littoral results from regular pulsations of flow and leads to fluctuations of the useful weighted area. It does not fit into the concept of fluvial continuum.

4. Rivers with a Mediterranean hydrological regime

Rivers of the Mediterranean littoral are characterized by the magnitude of the flow from autumn to the beginning of spring and the severity of the summer drying out. In the Argens, in Provence, the flow may range from more than 40 m^3/s in April to less than 2 m^3/s in August. In the lower Tavignano, in Corsica, it ranges from 26 m^3/s in December to 0.6 m^3/s in August. The littoral of rivers is mobile, with a pioneer vegetation in the flooded zone each year. The annual range of temperatures is about 15 to 20°C, with winter minima of 4-5°C in the lower Tavignano and 8-12°C in the streams of Provence.

4.1. Settlements of permanent rivers

The opposition between rhithron and potamon is accentuated in rivers with an oceanic pluvial or nival hydrological regime. The fauna of the low valleys is made up of:

— Mono-or polyvoltin species of winter and spring, belonging to the rhithron. These develop in summer at higher altitudes.
— Polyvoltin species with continuous development.
— Stenothermic species of warm waters, with summer development.

The important fact is the continuous settlement of the low valleys with a succession from a summer fauna to a winter fauna on the basis of species present in all seasons, unlike the settlement of the high valleys (Table 15). In the lower Tavignano, below 250 m altitude, 52% of species develop at the beginning of spring to the end of summer, 31% in summer, and 17% only in winter and spring. These last subsist even at 300 m altitude, but they represent only about 2% of individuals. It is slow current species that regress most rapidly with altitude: 20 species of Coleoptera and 6 of Hemiptera Heteroptera below 400 m, respectively 14 and 5 between 400 and 1000 m, and respectively 4 and 1 between 1000 and 1350 m altitude.

Table 15. Percentages of invertebrate species present in spring, summer, or both seasons in the Tavignano (Corsica), as a function of altitude (J. Giudicelli, 1968)

Altitude	April-May only	July-August only	April-May and July-August
< 400 m	17%	31%	52%
400–1000 m	18%	44%	38%
1000–1350 m	7.5%	57%	35.5%
1350–1720 m	2%	79%	19%

4.2. Temporary streams

The accentuation of summer low water may go as far as the temporary drying out of the lower course, or even the entire course in some small streams.

Drying out in the summer eliminates (1) species of running water whose biological cycle is longer than one year when they cannot find shelter in a sub-flow and (2) those whose emergence (for insects that become flying adults) or reproductive period occurs in summer.

Apart from these absences, the species found in winter and the beginning of spring in temporary courses do not differ appreciably from those of similar calm, permanent waters, with a predominance of insects. Their success in a temporary environment is due to the synchronization of their biological cycle with the hydrological cycle. The receding phase in spring is marked by progressive reduction in the number of species, due particularly to the emergence of monovoltin insects of the winter with flying adults. The reduction is accentuated in May-June, when the current has become very slow, while the surviving species concentrate in the residual ponds. These ponds are colonized by algae, notably optional heterotrophs such as Euglena, and invertebrates such as Mollusca Gasteropoda, Crustaceae Copepoda, Cladocera, *Asellus aquaticus*, *Gammarus pulex*, Heteroptera, and Coleoptera Dytiscidae.

When the stream flows over a rocky substrate or impermeable sediments (such as the Suberoque stream, in Provence), the survival of organisms is related to their potential for passing from slowed-down states of life to desiccation (Table 16). These phases of slow life are not at all obligatory in the course of the biological cycle. The species of temporary waters are simply the usual species of permanent waters, with very wide ecological tolerances, and swimming species that habitually colonize zones of slow current, the potamal. When the water table is close to the surface and coarse alluvia (Destel stream in Provence, France), some species burrow into the hyporheic environment and remain active while the surface dries out.

In autumn, when the water rises again, the temporary waters are recolonized by:

— the resumption of activity of species that remained in place in a slow state of life;

— the return to the surface of species that survived in the ground waters;

Table 16. Modes of survival during drying out of streams in Provence (P. Legier, 1979)

	Examples
States of slowed-down life in egg stage	*Nemoura cinerea* (Plecoptera), various Diptera Chronomidae and Culicidae
in larval stage	Crustacea Copepoda, Ostracoda, *Capnia bifrons* (Plecoptera), Diptera Ceratopogonidae
in adult stage	Mollusca Gasteropoda, most Coleoptera and Heteroptera of temporary waters
Activity maintained in ground waters	*Dugesia tigrina* and *D. subtentaculata* (Turbellaria), young *Ancylus* (Mollusca) Cyclopides, *Gammarus pulex, Asellus aquaticus* (Crustaceae), Hydracarids *Oulimnius, Bidessus* (Coleoptera), *Caenis macrura, Habrophlebia* sp. (Ephemeroptera)

— egg-laying by flying adults from nearby permanent waters;
— contribution by drift of organisms from the upper permanent course.

The similarity during winter and early spring between permanent sectors with slow current and temporary sectors is linked to the duration of the flooded phase. In the Peruÿ stream (Provence), which dries out between May and December, only 18% of species are common to the permanent and temporary sectors. Conversely, in the Suberoque and Destel streams, which flow for 7 to 9 months, 33 to 40% of species of the winter and early spring are common to both sectors.

4.3. Conclusions

Rivers of the Mediterranean region are characterized by a mobile littoral that results from significant variations in the width of the bed occupied during high water and summer low water. The bed of these rivers is similar for the most part to the alluvial plain. The alluvial plain itself has regular fluctuations in flow so that Mediterranean rivers cannot be integrated into the concept of fluvial continuum.

The relatively high winter and spring temperatures in the Mediterranean region are probably at the origin of the continuous biota—between winter-spring and summer—of water courses in the low valleys. This continuous biota disappears at higher altitudes.

When the flow becomes temporary, there are phases of cyclical destruction of organisms followed by autumnal phases of recolonization by migration from the sub-flow, drift of organisms from the upper course, and egg-laying by flying adults of the nearby permanent water courses.

13

Ecological Impacts of Development of Water Courses

Water is an element essential to the existence of living things. It also conditions human life, and humans are always striving to exploit its use to the maximum. Before the development of railways and road networks, the major cities were built close to the sea or near navigable rivers, on the major maritime and river routes. Up to the 19th century, the major economic alliances were determined on the basis of water routes. At the end of the 18th century in France, for example, the only towns with a population of more than 50,000 that were not near sea ports were located on navigable rivers—Paris (between the confluence of the Seine with the Marne and the Oise), Lyon (at the confluence of the Rhone and the Saone), and Strasbourg (on the Rhine).

The demand for water for consumption, irrigation of land, and industry is high. In the major cities, about 0.5 to 1 m^3 per day per inhabitant is required.

To meet all these needs, people have from time immemorial joined together to build hydraulic works for mutual use—aqueducts, canals, dams, modifications of water courses to facilitate navigation, and structures to protect them from floods. The first such projects date from 3000 BC in the Middle East, for farmland irrigation. The Ezechias aqueduct in Jerusalem was built in 725 BC, and at the beginning of this era, six aqueducts supplied Rome with around 1 million m^3 of water per day. In China, the Imperial Canal was built gradually from the 5th century BC to the 13th century, between Han K'Eou and Peking, over a length of 1780 km. Major developments of the 20th century included the construction of large dams for hydroelectric power generation. There were also human interventions on watersheds and extraction of gravel and other materials from riverbeds for use in construction.

The ecological impact of these developments differs according to their nature and their location on the water course, whether rhithral or potamal.

1. The Lot: a river subject to multiple developments

The Lot, a tributary on the right bank of the Garonne that springs from the Massif Central, is 491 km long and has a pluvio-nival hydrological regime.

Man-made constructions and activities along this river are of four types:

— embankments designed to render the Lot navigable, in the potamal;
— dams with a variable useful capacity in the upper valley (rhithral) or low capacity dams within the current, in the potamal (Fig. 50);
— extraction of water for irrigation, estimated at 5-6 m^3/s in summer;
— extraction of material, mainly gravel, from the bed.

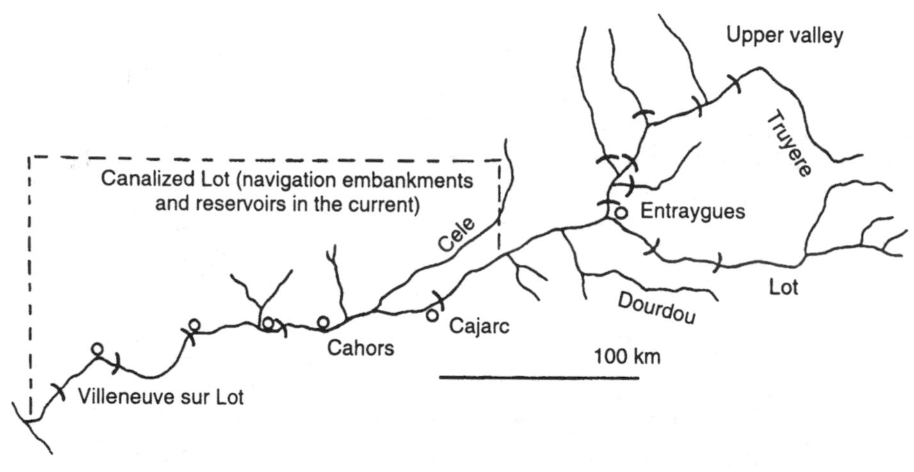

Fig. 50. Hydroelectric constructions on the Lot basin

The first developments to make the Lot navigable began in the 14th century and continued until the 19th century. From downstream of Entraygues to the confluence with the Garonne, about 60 embankments limited the reaches from a depth of 1.1 to 2.5 m. The dry bed at low water disappeared, and the submerged section was greater in relation to the summer flows. A current of 0.25 m/s appeared to be the critical speed between the phenomena of erosion and sedimentation, and this speed was reached only for flows of 30 m^3/s. Sediments were consequently deposited in the canalized reaches during the summer (Fig. 51). The hydroelectric installations in the upper valley began in 1930 on the Truyère, a tributary on the right bank of the Lot, but they were really only developed from 1945 to 1962. Seven reservoirs on the Truyère have a total storage capacity of 538×10^6 m^3. These reservoir-barrages, with a large capacity, store water from winter to spring, and the waters of deeper layers help cool the waters of the Truyère downstream during the summer. The two reservoirs on the Lot, on the other hand, with a small capacity (22.1×10^6 m^3) and low water retention time, help to heat the waters of the Lot downstream.

The reservoirs are managed by the sluice method: i.e., massive releases of water during peak hours of electricity consumption.

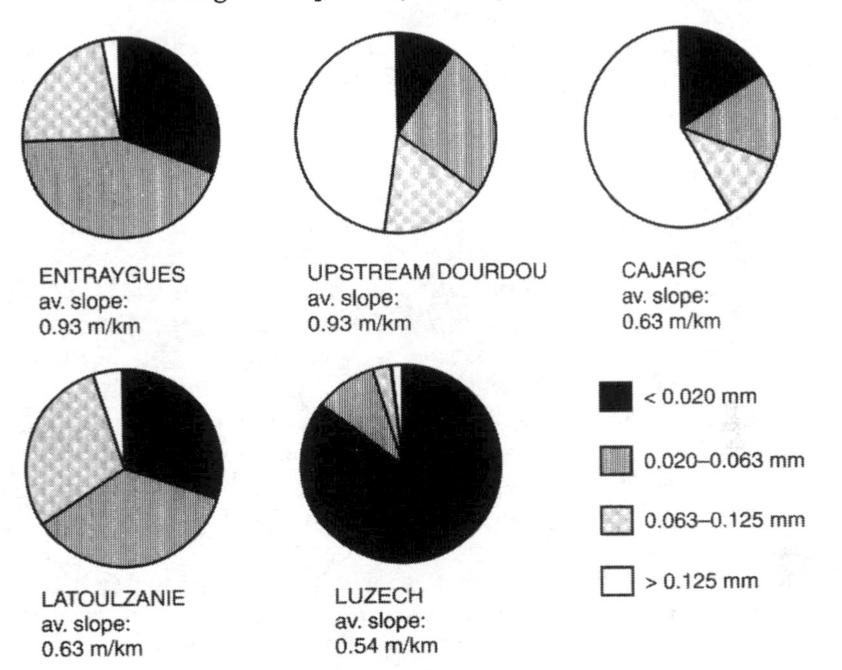

Fig. 51. Granulometry of fine sediments of the canalized Lot (J.M. Bordes, J.Cl. Lucchetta and M. Rochard, 1973)

Downstream of the Entraygues, the reservoirs have no useful capacity: built in the current, they simply ensure the levelling out required for the functioning of hydroelectric plants.

By means of a constant release from the barrage-reservoirs of the Truyère, a minimum flow of 6 m^3/s is maintained in the Lot during low water. Calculated in monthly averages, low water flows are reinforced by around 8% compared to natural flows, but this takes into account the natural flows (40 to 60 m^3/s, 4 to 6 h per day). For 16 to 18 h per day, the summer flow affected is less than the natural flow. These fluctuations in flow (which may reach a ratio of 1 to 10 in a day) generate a shock wave that propagates itself more rapidly than the flow itself. The result is a resuspension of clays and silt, which do not have time to become redeposited between two pulsations—hence the permanent turbidity of the tributary. The reservoirs in the current play a decanting role at low flows: that of the Cajarc, for example, constitutes a sedimentation zone for flows lower than 30 m^3/s, and of resuspension above that flow.

Extraction of gravel has increased greatly since 1945 (a total of 35 installations and 1 million m^3 were extracted in 1972). Gravel extraction contributes to the deepening of the river bed, and it returns the finer sediments into suspension.

The navigation embankments constructed downstream of the Entraygues have had the consequence of extending the potamal upstream. The hierarchy of ecological factors is modified in summer, during low water. The limiting factor is thus no longer the current, as much for benthos as for fish, or plankton transit time.

1.1. Phytoplankton

A development model of phytoplankton of the Lot was constructed to simulate its dynamics as a function of water transit time, luminosity, temperature, and nutrients—nitrates, phosphates, and silica (Chapter 7, sections 1 and 2). This model was validated for a constant flow of 20 m^3/s. However, the mode of management of reservoirs leads to highly variable flows during the day, and the water quality of the Lot differs according to whether it flows from the reservoirs of the Lot itself or those of its tributary, the Truyère. The same is true of algae of the phytoplankton. It has been seen (in Chapter 7) that the phytoplankton of the Lot appears as a succession of communities, each having its own history as a function of the turbidity, flow variations, temperature, exposure to sunlight of each water mass, and the reseeding it has received (Fig. 52). The flowing water appears not as an upstream-downstream continuum, but as a series of microhabitats that succeed one another within the current.

1.2. Benthos

From the source to the confluence with the Truyère (km 179), the benthos of the Lot corresponds to the classic crenon-rhithron succession. At the confluence, the temperature of Lot waters is 23-24°C in summer, and that of the Truyère 16.5°C at most. The mean flow of the Truyère being greater than that of the Lot in summer (because of the release of reservoir waters), the temperature of the Lot is lower downstream of Entraygues (21°C in August) than upstream. *Tipula montium* (Diptera) imagos appear in May-June upstream of Entraygues and only in July-August downstream.

From the Lot-Truyère confluence to the canalized Lot, the slope diminishes. It is a transition zone in which the species of the rhithral gradually disappear—Oligochaeta Naididae, Mollusca *Ancylus fluviatilis*, 34 taxa of Ephemeroptera out of 39 (in the entire river), 38 taxa of Trichoptera out of 43. The aprons are colonized by rheophilous macrophytes (*Ranunculus*, Bryophytes). In the reaches downstream with movable bottoms, the first invertebrates of the potamal appear (Oligochaeta Tubificidae, Cladocera, Odonata, Diptera Chironominae).

From the first navigation embankments, the summer temperature increases (25°C in August). The macrophytes are potamophile species, but their development is limited by the steep banks. The vegetation of *Potamogeton, Polygonum, Typha*, and other genera hardly appear, except upstream of the reservoirs in the current and at some points of the lower course.

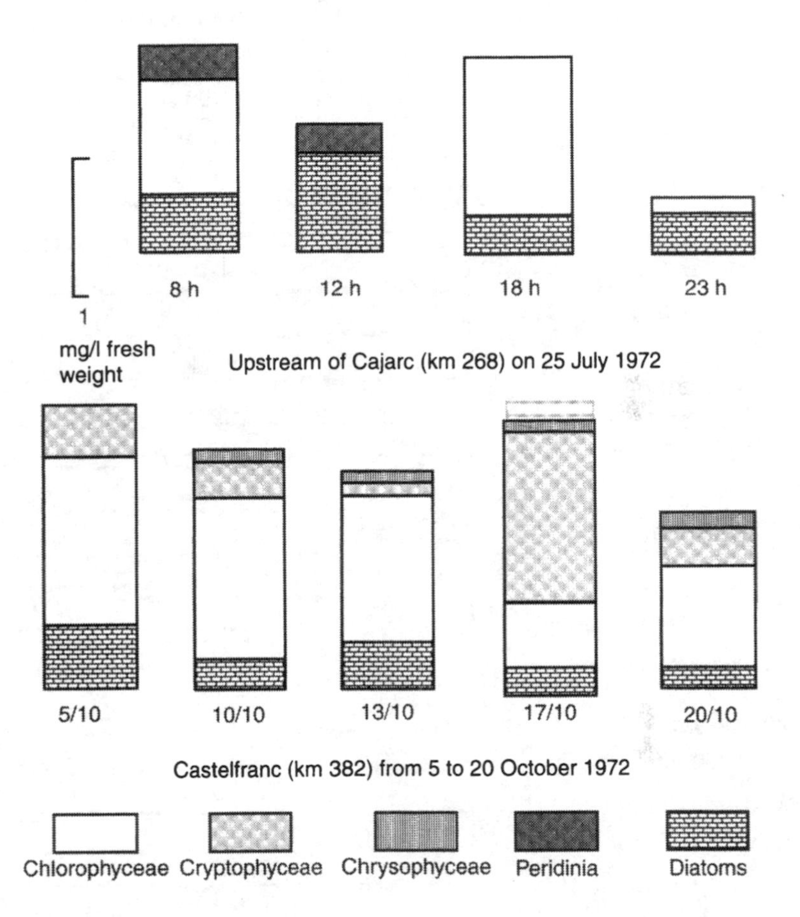

Fig. 52. Occasional variations of the biomass and components of the phytoplankton in the Lot (A. Dauta, 1975)

The depth of reaches and turbidity limit the penetration of light and the development of benthic algae. On the entire canalized Lot, the species richness of invertebrates is poorer than that of a natural potamon. While the species richness of molluscs is higher on a potamal than on a rhithral, only 7 species (out of 17 for the entire Lot) colonize the canalized Lot, 12 species of Oligochaeta out of 32, 5 species of Hydracarids out of 31, and 2 species of Coleoptera out of 19.

1.3. Fish

Some 28 species of fish, of which 4 are migratory, have been recorded on the Lot river (Table 17). Salmon, which was abundant in the 18th century, had already disappeared when the first hydroelectric dams were constructed.

Table 17. Distribution of fish species in the canalized Lot before 1932 and in 1975 (J.N. Tourenq and F. Dauba, 1978)

	Km 200	Km 300	Km 400
Phoxinus phoxinus			
Barbus barbus			
Leuciscus cephalus			
Leuciscus leuciscus			
Cobitis barbatula			
Chondrostoma nasus			
Chondrostoma toxostoma			
Cyprinus carpio			
Tinca tinca			
Alburnus alburnus			
Lepomis gibbosus			
Abramis brama			
Species introduced in reservoirs in 1948 *Esox lucius*			
Stizostedion luciperca			
Micropterus salmoides			
Ictalurus melas			
Distribution in 1975 ▬	Distribution before 1932 —		

It bred in the small tributaries of the left bank of the Lot, upstream of Cahors. Before hydroelectric development, the salmonid zone extended from the upper watershed to the Lot-Truyère confluence. The barbel zone began downstream of Entraygues.

The ecological impact of only the hydroelectric works is known, the navigation embankments having been built before the first fish census. That impact manifests itself differently according to the various levels of the Lot.

The reservoirs in the current are an obstacle to the upstream migration of fish. Mullet, shad, and sea lamprey do not cross the Temple reservoir (km 480), which was built in 1948. The eel is in regression from the Temple reservoir onwards, rare at the Lot-Truyère confluence, and absent further upstream.

The Castelnau reservoir (km 130) and the Golinhac reservoir (km 160), at the rhithral, have a low capacity and contribute to the heating of Lot waters in summer, up to Entraygues. Nase, gudgeon, and minnow have disappeared. The trout, chub, and barbel populations have remained stable. The small tributary streams serve as shelters for trout in summer and as spawning areas in autumn and winter. The introduced species—pike, roach, dace, carp, and tench—maintain their numbers, while the pikeperch is clearly increasing. Upstream of Entraygues, barbel tends to dominate.

The stretch downstream of the Lot-Truyère confluence used to be a barbel zone. Gudgeon, barbel, nase, and dace are in regression, while trout proliferates in the waters, which have become colder in summer, and occupies the place that dace held before 1932. The populations of minnow, chub, loach, pikeperch, and European brook lamprey are stable, while pike is not abundant. In the absence of reproduction, roach, carp, and tench populations can only be maintained by farming.

On the canalized section of the Lot, censuses before 1932 showed the classic potamon succession, barbel and bream zones. The distribution of six species in 1975 was identical to that of 1932 (along the whole of the potamon group): roach, gudgeon, perch, rudd, toxostome, and European brook lamprey.

Trout, which was rare before 1932, hardly survives except in the tributary streams.

Seven species, including barbel, minnow, chub, and dace, which used to extend over the entire canalized Lot, are now confined to the upper part. Five others, on the other hand, have extended their distribution upstream: carp, bream, bleak, and sun perch among them. Species introduced into the reservoirs since 1948 colonize either the entire canalized Lot (pike) or just the lower Lot (pikeperch, black bass, catfish).

2. Ecological impacts of regulated flows on the rhithron: the Verdon

The Verdon, a tributary on the left bank of the Durance, in Provence, is 155 km long and has a pluvio-nival hydrological regime. The average slope varies from 2.1% in the upper valley to 0.39% in the lower valley. The natural course is a rhithral.

Five reservoirs were built from 1947 to 1974 (Fig. 53). These are barrage-reservoirs with a large capacity and average depth varying from 50 m (at Quinson) to 95 m (at Castillon). The reservoirs are managed by means of sluices. In the middle Verdon, between the Chaudanne and Sainte Croix (Verdon gorges) reservoirs, the flow varies from 0.5 to 45 m^3/s during the day. Conversely, in the lower Verdon, downstream of the

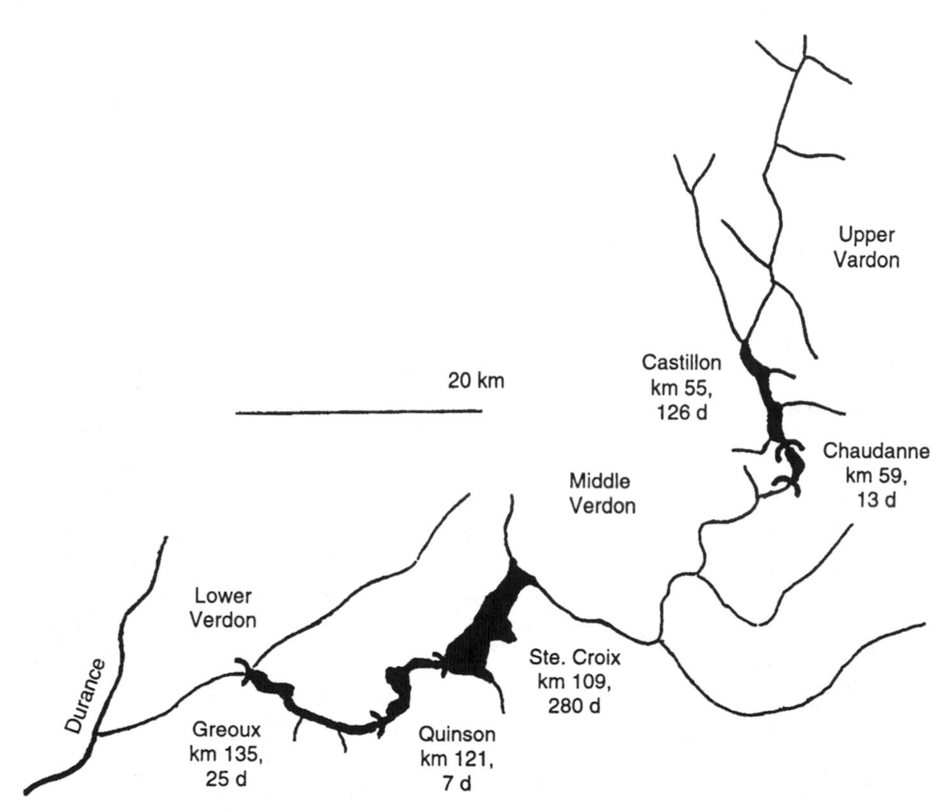

Fig. 53. Reservoirs of the Verdon basin. Distance from the source to the reservoirs and average transit time of water in days (A. Gregoire, 1982).

Greoux reservoir, the flow is slow (1 m^3/s) but constant over 9 km, up to the restitution of drifting waters. The speed of the current is generally less than 0.30 m^3/s.

The reservoir water points are close to the bottom, where the temperature in summer is between 10 and 12°C. The result is a cooling of waters from the middle and lower Verdon, more so than is recorded in the Lot downstream of Entraygues. At the foot of the Chaudanne reservoir, the average temperature of the Verdon is 11°C in August. It reaches 16°C at the entry of the Sainte Croix reservoir, after a run of 38 km. In the lower Verdon, the temperature at the foot of the Greoux reservoir is also 11°C, but the low flow and the summer temperature of the air leads to a rise of 7°C over just 9 km.

2.1. Benthos

The impact of regulated flows is different in the middle Verdon, which is subject to a sluice regime and daily leaching of the substrate, and the lower Verdon, where the slow, regular flow allows sedimentation of fine particles.

Plankton from the reservoirs disappears rapidly (95% in the first kilometre downstream of Chaudanne). The benthic algae are essentially Diatoms. Their species richness is low, as is their density (20×10^3 cells/cm^2 downstream of the Chaudanne reservoir, 27 to 77×10^4 upstream of Castillon). From the time the impact of the sluices is buffered, the density, but not the species richness, increases again (83 to 160×10^4 cells/cm^2). The colonization of the substrate is greater on the lower Verdon, which is not subject to the sluice effect.

Sluice action maintains the diversity of microhabitats on the rhithral and possibilities of colonization of the hyporheic environment (the interstices are not sealed). Invertebrates can thus migrate towards protected microhabitats or into the hyporheic environments during the working of the sluices. The drift is nevertheless powerful at the beginning of sluice action.

Immediately downstream of the Chaudanne reservoir, the river is colonized by rheophile larvae of insects. Six species are dominant: *Ecdyonurus forcipula* and *Rhitrogena aurantia* (Ephemeroptera), *Perla maxima* and *Amphinemura sulcicollis* (Plecoptera), *Hydropsyche angustipennis* (Trichoptera), and *Limnius* sp. (Coleoptera). Amphipoda *Gammarus pulex* is abundant, and there are also Entomostraceae Copepoda and Ostracoda known to be from the hyporheic environment. The impact of sluices is buffered within 2 km, and subsequently there is a tendency towards a stenothermic fauna of cold waters of the natural stream (Table 18).

On the lower Verdon, the low and regular flow favours vegetation colonized by a limnophile fauna composed of Hirudinea, Odonata, Ephemeroptera such as *Potamanthus luteus*, and Coleoptera Dytiscidae. The diversity of microhabitats is reduced by the sealing of the substrate. There is a trend towards a fauna of the potamon type until the waters are restored at the Greoux reservoir, at 9 km downstream. Rheophile invertebrates such as Ephemeroptera Heptageniidae then reappear.

2.2. Fish

Trout is the dominant species on the undeveloped sector of the upper Verdon (Table 19). It is also dominant on the middle Verdon, but with numbers reduced to about 80% compared to the upstream sector. Four other fish species are found in these reaches, all less abundant: chub, barbel, blageon, and eel. Some species of the Sainte-Croix reservoir migrate a little upstream of this stretch. The reduction in numbers is due to a small, permanently submerged section, together with a shaking up of the bottom sediments by sluicing effect, drying out of spawning areas, and limited food sources.

Table 18. Distribution of Ephemeroptera on the Verdon (A. Gregoire, 1982)

	Upper Verdon	Ca Ch	Middle Verdon	SC Q, G	Lower Verdon
Baetis melalonyx	—				
Baetis alpinus	—				
Epeorum torrentium	—		—		
Baetis vernus	—		—		—
Paralepto submarginata	—		—		—
Baetis sinaicus	—		—		
Rhitrogena aurantiaca	—		—		—
Ecdyonurus forcipula	—		—		
Baetis minutus	—		—		
Oligoneuriella rhenana			—		
Baetis rhodani	—		—		—
Caenis moesta	—		—		—
Cloeon simile	—		—		—
Ephemerella ignita	—		—		—
Baetis lutheri			—		—
Centroptilum luteolum			—		—
Ephemera danica			—		—
Potamanthus luteus					—
Heptagenia sulphurea					—

Ca, Ch, reservoirs at Castillon and Chaudanne.
SC, Q, G, reservoirs at Sainte Croix, Quinson and Geroux.

Species of the middle Verdon are found downstream of the Greoux reservoir. However, trout numbers decrease in favour of the nase. Minnow, gudgeon, chub, stone loach, and blageon are also present.

3. Conclusions

In this chapter, we have seen the basics of the development of water courses between the Lot and the Verdon. The ecological impacts of such development depend on its nature and its location on the water course (rhithral or potamal).

Developments in the potamal are the oldest and were designed as protection against the effects of floods (dykes) and to facilitate navigation (embankments). Dykes, by reducing the area that is flooded, increase the violence of floods and thus the churning up of the substrate. Navigation embankments, by raising the water level, keep the bed submerged even at

Table 19. Distribution of fish species on the Verdon (A. Gregoire, 1982)

	Upper Verdon	Ca Ch	Middle Verdon	SC Q, G	Lower Verdon
Salmo trutta	—		—		—
Telestes soufia	—		— —		—
Esox lucius	—		—		—
Leuciscus cephalus			—		—
Anguilla anguilla			—		—
Rutilus rutilus			—		—
Barbus barbus			— —		— —
Chondrostoma toxostoma			—		—
Zingel asper			—		
Chondrostoma nasus					—
Phoxinus phoxinus					—
Cottus gobio					—
Cobitis barbatula					—
Perca fluviatilis					—
Barbus meridionalis					—
Gobio gobio					—
Spirlinus bipunctatus					—

Ca, Ch, reservoirs at Castillon and Chaudanne.
SC, QG, reservoirs at Sainte Croix, Quinson and Geroux.

low water. The submerged section of the water course is thus too large in relation to the flow, and the increased water transit time at the potamal favours the development of phytoplankton. This is the beginning of the process of eutrophication (see Chapter 14).

The depth of reaches and the slope of banks limit the development of macrophytes of the potamal and littoral fauna. Species richness is lower on a developed potamal such as the Lot than on a natural potamal. The heating of waters in summer is due to the large surface area of the water body between two successive embankments as well as the long water transit time. The cyprinid zones extend upstream.

Twentieth century developments were essentially aimed at hydroelectric production and involve mostly the rhithral (except for reservoirs in the current). Large-capacity reservoirs, with a long water retention time, store cold waters during floods—between autumn and spring in temperate climates. When the water supply points are located near the bottom, the release of cold waters leads to the cooling of river waters and the downstream extension of salmonids. When the water supply points are located close to the surface or on reservoirs with a short transit time, their increasing temperature leads to the heating of river waters (as in the Lot upstream of Entraygues).

Reservoirs are a hindrance to the circulation of fish and especially to the upstream journey of migratory species. On the Lot, for example, only the eel manages to cross the first reservoirs from downstream.

When reservoir waters are diverted (into irrigation canals or diversion canals towards a hydroelectric plant), only a very slow flow remains in some sections of the river (called reserve flow). The submerged section is reduced and the current is slow. Fine sediments are deposited and seal up the substrate and the hyporheic environment, reducing the diversity of microhabitats as well as species richness. There is a trend towards a settlement of the potamon type, at the expense of rheophile species.

The most widespread mode of reservoir management in France is the sluice system, involving the temporary release of water during peak hours of electricity consumption. Downstream of the reservoirs, for 4 to 6 h per day, the flow may go from 6 to 60 m^3/s on the Lot and 0.5 to 45 m^3/s on the middle Verdon. In the rhithral, the diversity of microhabitats is maintained by the regular sweeping away of the finest materials (clays and silt). Only the most rheophile species or those that can migrate into the hyporheic environment can maintain themselves immediately downstream of the reservoirs. The low density of the benthos is linked to the significant drift at the beginning of sluicing and to the constant recirculation of the bottom sediments. The low numbers of fish are also due to the recirculation of bottom sediments and the churning up and periodic drying out of spawning areas. Further downstream of the sluice gates, their effect is buffered. The inputs of tributaries help stabilize the flow, and there is a gradual return to a natural rhithron.

14

From Eutrophication to Trophic Pollution

The word *eutrophication* signifies an increase in autochthonous plant production (macrophytes, periphyton, and phytoplankton) correlated with a high level of nutrients in the water, particularly nitrogen and phosphorus salts. Eutrophication corresponds to the increase of energy potential of the environment, which has repercussions on the entire ecosystem across its food webs, in running as well as stagnant waters.

Eutrophy is the opposite of oligotrophy, which is characterized by a low level of nutrients and plant production, while mesotrophy is an intermediate stage.

Carbonic anhydride (CO_2) needed for plant photosynthesis is dissolved at the water-atmosphere interface. Nitrogenous compounds (ammoniac, nitrites, nitrates) originate from dissolved nitrogen fixed by heterocystate Cyanophyceae, autotrophic or heterotrophic bacteria, and, for the most part, the oxidation of nitrogenous compounds from the watershed. Phosphorus is a rare element of the mineral world. In natural waters, it is most often the limiting factor of plant production. In lakes there is a correlation between phytoplanktonic biomass and phosphorus concentration.

The role of phosphorus in lakes was demonstrated by D.W. Schindler (1977) on a Canadian lake divided into two parts. Soluble phosphorus was added to one of the two halves of the lake. The result was rapid eutrophication, and the flow of other nutrients (CO_2, N/HN_3, NO_2, and NO_3) increased at the water-atmosphere interface.

In the second half of the 20th century, development of collective drainage and organic wastes, especially from agro-food industries, contributed to the release of nutrients into water bodies. A large part of the phosphorus wastes came from phosphate fertilizers and polyphosphates from detergents, which contributed to the eutrophication of waters.

Lake Leman (in France) was oligo-mesotrophic in 1960, with N and P values close to natural values. It became meso-eutrophic when phosphorus concentrations exceeded 30 μg/l.

Eutrophication models were designed for lakes, notably the Vollenweider model. They highlighted the relationship between chlorophyll concentration and phosphorus concentration. The amount of phosphorus is established by taking into account its concentration per unit of surface area of the lake, its sedimentation on the bottom, average depth, and water transit time. The last factor varies from some weeks for a pond to a few hundred years for very deep lakes.

In such models, empirical or statistical, the chlorophyll-phosphorus correlation is practically linear up to phosphorus concentrations of 100 mg/l. This signifies that beyond that level, phosphorus ceases to be the limiting factor of planktonic production. Thus, dynamic models are used taking into account nutrient concentrations (including silica needed by Diatoms) and factors such as light, temperature, phosphorus cycling time, and browsing of phytoplankton by zooplankton.

1. Eutrophication in running waters

Apart from known factors of eutrophication in lakes, the current intervenes in rivers as a factor of erosion, limiting the development of macrophytes and phytobenthos, and by short water transit times (Chapter 7, section 1.1). The short transit times lead to an under-exploitation of nutrients by phytoplankton and of the capacity of zooplankton to regulate the phytoplankton by browsing.

Much more than in lakes with a high phosphorus concentration, each water course is unique, characterized by its current, transit time, temperature, insolation, and zonation (rhithral or potamal).

1.1. Eutrophication of the Upper Aveyron

The Aveyron is a tributary of the Tarn, to the south of the Massif Central in France. The upper valley extends from the source (altitude 710 m) to downstream of Rodez (490 m), over a length of 60 km. The summer flow is about 1 m^3/s. The average slope (0.3%) is that of a rhithral (which also reflects the annual average temperature, lower than 15°C). It is a salmonid river. However, embankments have created many reaches that slow down the current, so that substrates of pebbles and silt are located side by side. Benthic invertebrates represent this ambiguity: molluscs such as *Ancylus fluviatilis* are found on the entire upper Aveyron, but, at some points, so are lamellibranch molluscs *Unionidae* and *Sphaeridae*.

Thirty per cent of the nitrogen and 75% of the phosphorus in the river is due to occasional inputs near towns (population 55,000, of which more than half is at Rodez). The diffused inputs (erosion of soil and farm fertilizers) essentially consist of nitrogen. If just the nitrogen and phosphorus concentrations (Fig. 54) were considered, the upper Aveyron would be thought a highly eutrophic river.

However, phytoplankton development is low: 1.7 mg/l of chlorophyll, on average, and 7.6 mg/l at most in summer. Conversely, benthic algae density is high: from 32 to 150 mg/m^2 of chlorophyll in summer. The

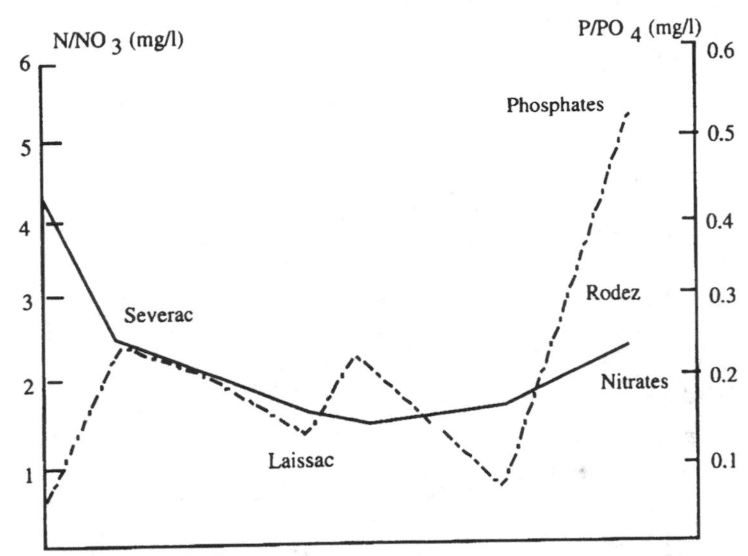

Fig. 54. Upstream-downstream trend of concentrations of nitrates and phosphates on the upper Aveyron. Averages from July to September (Chauvet, Prat and Tourenq, 1983).

percentage of substrate cover varies from 10 to 60%. Chlorophyll concentrations correspond to phytobenthos of a eutrophic environment at a low current. Nevertheless, there seems to be no correlation between algal growth and nutrient concentration: maximum phytobenthos density can only account for 1% of nitrate loss, at most. It is bacterial denitrification of the phytobenthos and the substrate that seems to be behind the nitrate losses observed. The nutrients, which are in excess, are not the limiting factor for algal growth: current, insolation, and temperature remain the essential regulatory factors.

Macrophytes, absent from Severac to Laissac, occupy 30% of the substrate close to the source and 10 to 50% from Laissac downstream of Rodez. *Myriophyllum spicatum* is considered an indicator of eutrophication, as is *Nuphar lutea*. *Potamogeton trichoides* is particularly developed downstream of water purification plants.

The incidence of nutrient absorption in the water by macrophytes is low. Macrophytes essentially exploit the substrate water (in which phosphorus is trapped) through their root system, even in species that have a poor root system.

The presence of vegetation leads to a release of oxygen during the day, which leads to a slight oversaturation of the water.

1.2. Eutrophication of the Lot

The problem of phytoplankton growth in the Lot has already been discussed (Chapters 7 and 8). With only rare exceptions (Fig. 55), in August 1971

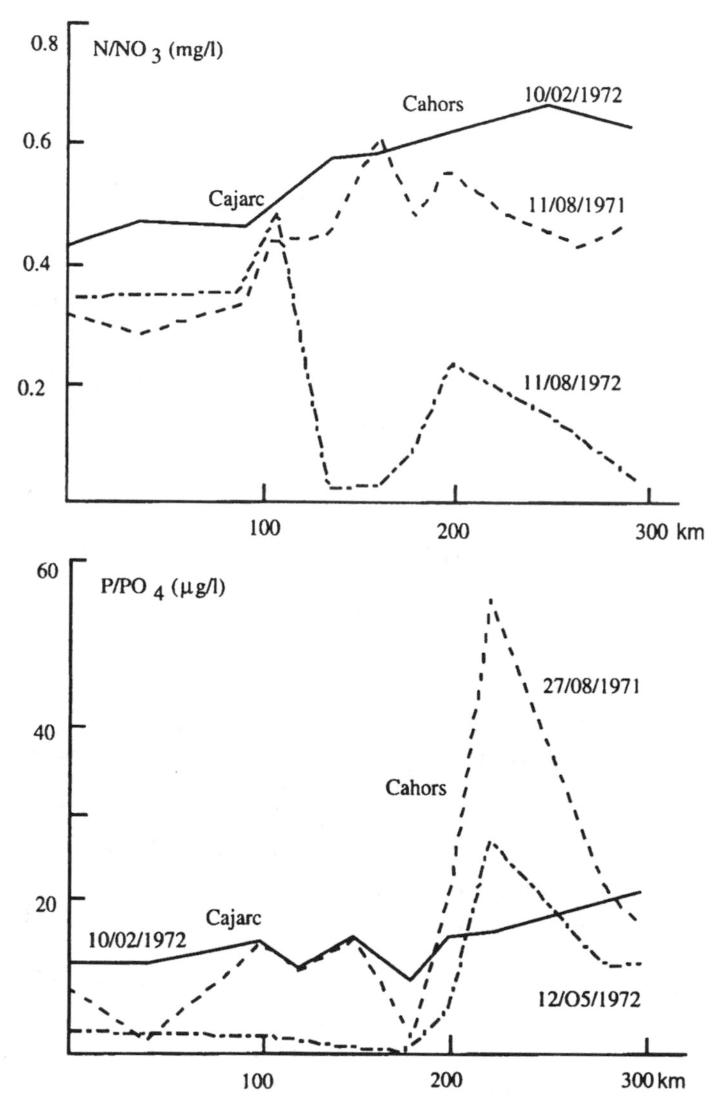

Fig. 55. Upstream-downstream trend of concentrations of nitrates and phosphates on the canalized Lot (J.M. Bordes, J.Cl. Lucchetta and M. Rochard, 1973)

(at a water transit time of 18 to 25 d for flows of 40 to 50 m^3/s), the summer transit time was between 60 and 85 d and no longer constituted a limiting factor for phytoplankton development, at least for Diatoms and Chlorophyceae.

Nutrient inputs are occasional (from towns) as well as diffuse (from farms). Nitrate and orthophosphate concentrations were higher in winter than in summer and increased quite regularly from upstream to downstream. From late spring to the autumn, there was a highly significant correlation between nutrient concentrations and phytoplankton biomass. The longitudinal evolution of algae was marked, in the ensuing water masses, by phases of bloom followed by exponential multiplication, and then decline more or less directly associated with the inputs of nutrients.

Nitrate and phosphate concentrations can fall to very low rates (Fig. 55: NO_3^- on 11 August 1972 and PO_4^- on 12 May 1972) or even to zero in some reservoirs. However, from 1973 to 1975, for phytoplankton biomass greater than that in 1972, the nutrient reserves were exhausted only once, in a reservoir on the lower course. This implies a seasonal or annual variability of limiting factors: transit time, temperature, turbidity, and nutrients.

The conditions under which Cyanophyceae blooms appear are present in flows lower than 10 m^3/s. The biodegradation of Cyanophyceae downstream of reservoirs enriches the nutrient content of the water and leads to the resumed growth of other algae.

Decline in silica concentration is associated with the development of Diatoms, which use silica for the constitution of their theca.

Apart from very localized points, downstream of towns the concentrations of nitrogen and dissolved phosphorus are limited, as is organic pollution. The nychthemeral variations of oxygen saturation are high only in the reservoirs in the current. However, in these reservoirs or in deep reaches, a chemical stratification may become established during low flows of summer, analogous to the stratification of lakes. This leads to significant oxygen deficits in the deep layers.

1.3. Eutrophication of the Charente

The Charente is a river of the Atlantic coast of France; it is 350 km long. It is a plains river (the altitude at the source is 250 m) flowing over chalky terrain (except for the first 40 km). The slope is gentle, especially in the last 100 km downstream of Cognac. Most of the course is a potamal, although navigation embankments raise the level of the water until Angoulême. The waters are of the pluvial oceanic regime. For an average flow of 60 m^3/s at the estuary, the flows may be lower than 5 m^3/s during severe drying out.

Five towns over the entire course have a population larger than 10,000 (with Angoulême at 87,000).

Nitrate and phosphate concentrations are much higher than those of the Lot—and close to those of the Aveyron (Fig. 56). However, unlike at Aveyron, transit time and high summer temperatures favour phytoplankton development. Over the first 120 km of the course, in the upper reaches, the high nitrate concentrations, and especially high phosphate concentrations, lead to phytoplankton development. Nutrients and chlorophyll fall gradually until Angoulême (where dissolved oxygen saturation is 105% during summer days).

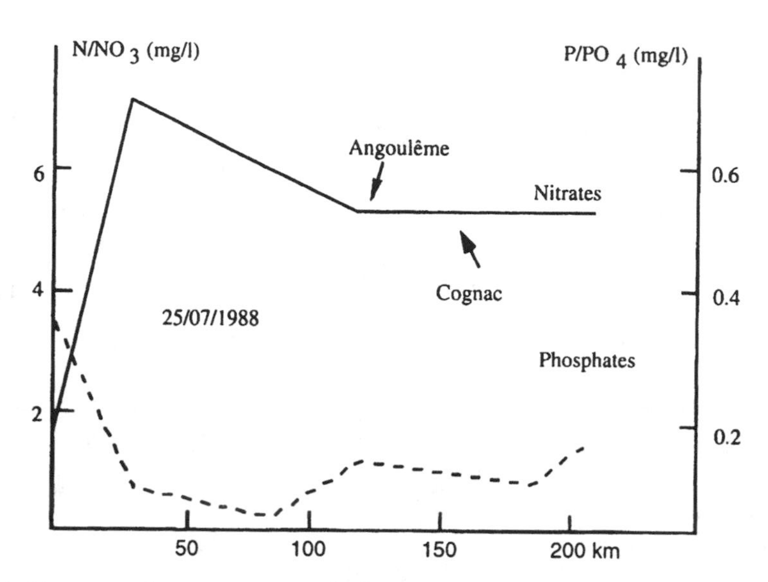

Fig. 56. Upstream-downstream trend of concentrations of nitrates and phosphates on the Charente, in summer (CEMAGREF document, Bordeaux, 1991)

Phosphate wastes from Angoulême lead to renewed algal growth, with a maximum of chlorophyll that juts out of line on the longitudinal profile in relation to the wastes. This interval corresponds to a transit time of 3-4 d and probably results from the shade cast by aquatic vegetation and riverside trees. Despite the input of oxygen by photosyn-thesis, the organic wastes at Angoulême cause an undersaturation of oxygen in the water. It reaches its maximum at 25 km downstream (average saturation 75%, 20 to 60% at the slowest flows). Oversaturation, i.e., a return to autotrophy, appears only at Cognac. The decline in phosphate concentrations, on the upper course and downstream of Angoulême, cannot be imputed only to their consumption by phytoplankton. Part of the particulate phosphorus is trapped by sediments (and released again during floods).

The phytoplankton of Charente is dominated by Diatoms and Chlorophyceae. Euglena, which are characteristic of organic pollution, appear at some places. Cyanophyceae are abundant only on a tributary, the Boutonne, which spills into the Charente close to the estuary, and sowing therefore occurs too late. Zooplankton is represented only by some ciliate protozoa and rotifers.

Macrophytes are abundant downstream of Angoulême: the depth of reaches appears to be the essential factor in the distribution of submerged helophyte species.

The Charente can be considered mesotrophic upstream of Angoulême, and eutrophic downstream, in terms of its chlorophyll concentrations as well as the presence of centric diatoms, which are characteristic of such environments.

1.4. Eutrophication of the Vire

The Vire is a river in Normandy that empties into the English Channel after a run of 122 km. It springs from the Vire-Carolles granite massif and in the last kilometres crosses pre-Cambrian terrain covered in permo-triasic deposits.

The river was developed in the 19th century to make it navigable. Out of 18 embankments, 12 are still in use and slow down water transit. Summer temperatures are between 20 and 25°C. Minimal flows are lower than 1 m^3/s.

Nutrient inputs come from cattle pastures and domestic and industrial wastes (especially from dairies). Nitrate concentrations are similar to those of the Charente, i.e., 2040 mg/l of N/NH_3–NO_2–NO_3 (240 to 7273 mg/l in 1990). Phosphate concentrations, on the other hand, are higher: 633 mg/l on average (287 to 2088 mg/l).

Neither nitrate nor phosphate exhaustion is observed during the summer on the canalized Vire. These two nutrients cannot therefore be considered limiting factors of phytoplankton development. Rather, the limiting factor is water transit time, and light and temperature are the amplifying factors.

Eutrophication is accelerated in summer for flows lower than 1.2 m^3/s. The opening of reaches, by reducing the depth of the Vire and transit time, limits phytoplankton development. Simulations of algal development were effected, from four species of Chlorophyceae and one species of Cyanophyceae (the Diatoms are marginal in summer), as a function of water transit time (severe low water flow of 0.5 m^3/s). The model takes into account the eutrophication of the canalized Vire, downstream of Pont Farcy (54 km), for a temperature of 20°C (Fig. 57).

The opening of all the reaches is equivalent to a transit time of 19 d. Chlorophyceae are dominant over the first 40 km. Cyanophyceae subsequently dominate, but their development is moderate. Maximum concentrations of chlorophyll are found near km 40. The rates of dissolved oxygen fall downstream of the wastes at Saint Lô, but rapidly revert to oversaturation.

The opening of just two reaches, the largest ones (from downstream of Saint Lô to km 54), increases transit time to 36 d. Cyanophyceae subsequently present blooms followed by collapses. Phases of anoxia (from the biodegradation of the Cyanophyceae) appear from km 40 onward.

When all the reaches are closed, transit time reaches 50 d. Cyanophyceae very quickly begin to dominate, and blooms appear even before km 40.

For a flow of 1.2 m^3/s, over all the closed reaches, transit time falls from 50 to 21 d. Again there is algal development analogous to that of

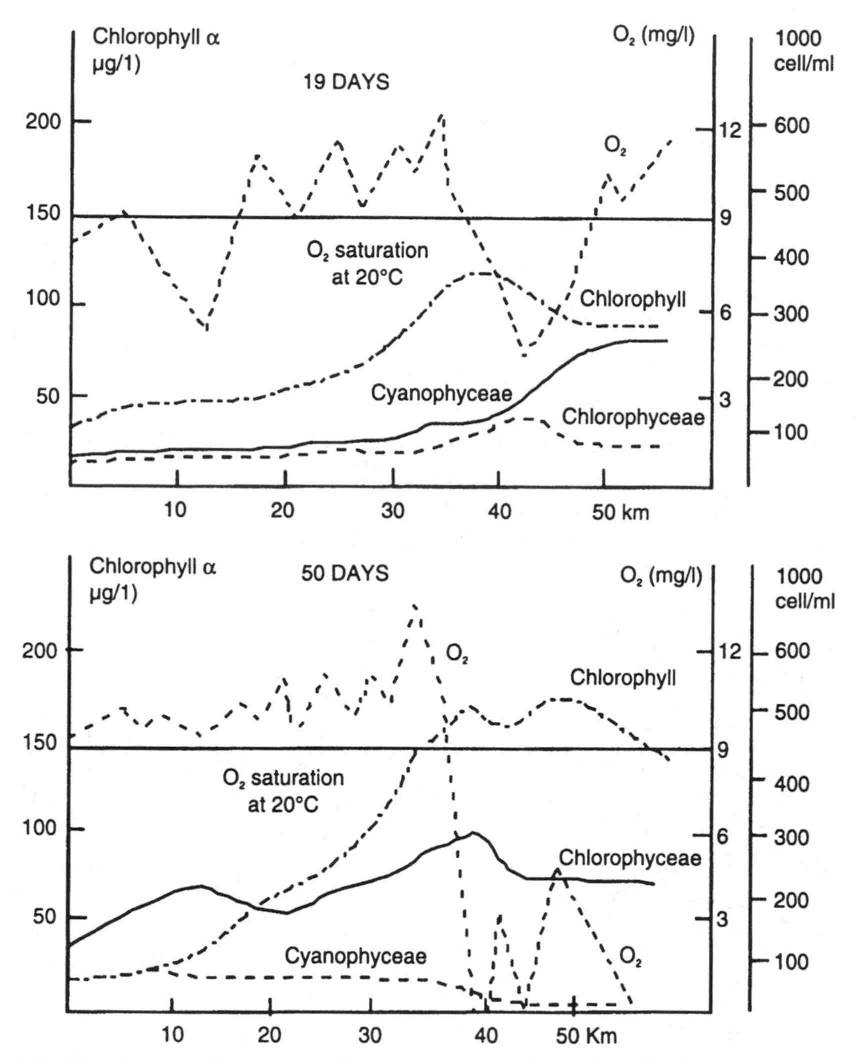

Fig. 57. Simulations of upstream-downstream trend of phytoplankton and oxygen concentration on the canalized Vire, as a function of transit time (F. Auscher, 1992, simplified)

empty reaches for a flow of 0.5 m^3/s. The model was validated by opening and closing the reaches.

The Vire presents the same phenomenon as the Lot: Cyanophyceae blooms correspond to long transit times. On the Saint Lô reach, in August, when there is a Cyanophyceae bloom, chlorophyll concentrations of 16,000 µg/l have been observed on the surface, 144 µg/l at 1 m depth, and 132 µg/l at 2 m depth.

The Vire eutrophication model shows great sensitivity to initial conditions. A reduction of nutrients up to 70%, at the beginning, reduces the development of Cyanophyceae. Beyond that, it is Chlorophyceae that regress. A reduction of 20% of initial phosphorus causes an appreciable regression of nitrogen-fixing Cyanophyceae.

Cyanophyceae is seeded from two tributary streams, and especially from the Dathee reservoir, upstream of the Vire. It is an irregular seeding, and Cyanophyceae blooms originate from different species depending on the year.

1.5. Conclusions

Throughout the upper Aveyron, the Lot, the Charente, and the Vire, the longitudinal profiles of nitric nitrogen concentrations are just about regular, without distinct peaks. The predominant nitrogen inputs are diffused, both urban and agricultural in origin. The longitudinal profiles of phosphates, on the other hand, show significant peaks downstream of towns (Rodez on the Aveyron, Cahors on the Lot, Angoulême and Cognac on the Charente, and Saint Lô on the Vire). Domestic urban and industrial wastes are predominant in the phosphorus inputs. The quantity of phosphorus due to domestic wastes is about 3.2 g per day per inhabitant, of which two thirds is accounted for by polyphosphate detergents. The commercial availability of this type of detergent has therefore tripled phosphorus inputs into water courses. Particulate phosphorus is trapped by sediments, but it may be recirculated during high water.

High phosphorus concentrations favour eutrophication, but the four examples given above show the complexity of the phenomenon in running waters. Each river is ultimately a unique case. Three factors regulate the development of phytoplankton: water transit time; depth; and nutrients (nitrogen and phosphorus, silica for Diatoms). To this should be added light and temperature, which are amplifying factors. Regulators and amplifiers differ from one river to another and depending on the season.

On the upper Aveyron, the nutrient concentrations could be factors of eutrophication; however, phytoplankton is poorly developed. This is in a rhithral where low summer temperatures and short transit times limit phytoplankton development. The density of benthic algae is a sign of eutrophication. However, the relationship between their growth and nutrient concentrations is weak.

On the Lot, transit time is a limiting factor for flows higher than 20 m^3/s. Below that, nutrients become a limiting regulatory factor. Their concentration is low when compared to those of the Charente and the Vire, and they may be exhausted from upstream from Cahors. Domestic and industrial (dairy) inputs from Cahors make the resumption of algal development possible. Cyanophyceae grow only in summer at flows lower than 10 m^3/s.

In the Charente, nitric nitrogen concentrations are nearly 10 times as high as those of the Lot. The same applies to phosphates downstream of

Angoulême in relation to those downstream of Cahors. The Charente, which is mesotrophic upstream of Angoulême, is truly eutrophic from Cognac onwards. The shade cast by macrophytes downstream of Angoulême helps limit algal growth. Phosphorus losses cannot be imputed only to consumption by plants. Particulate phosphorus, especially, is trapped by sediments.

Phytoplankton is dominated by Chlorophyceae and Diatoms. Seeding of Cyanophyceae is too slow to effectively modify the composition of phytoplankton and so promote blooms.

The ratio between nitrogen and phosphorus concentrations is close to 10/1 on the Charente, as on the Lot: i.e., nearly equivalent to the N/P ratio of plants.

On the Vire, the average N/P ratio is 3/1. Phosphorus is thus no longer a limiting factor (neither is nitrogen, given their concentrations). Transit time again becomes the regulating factor. Chlorophyceae and Cyanophyceae coexist when all the reaches are open (transit time 19 d over 54 km). The Cyanophyceae reach cellular densities of 500×10^3/ml for a transit time of 26 d. The successions of blooms and a biodegradation phase with anoxia appear from a transit time of 36 d onwards. Chlorophyceae are thus eliminated in the last kilometres. Anoxia affects the entire water column or just the deep layers when there is a thermal stratification isolating these layers from the surface waters.

Simulations on the Lot as well as on the Vire show the sensitivity of models to initial conditions. The potential for eutrophication, in each sector, is obviously related to the conditions that prevail in it, but also to the conditions of seeding, i.e., to phenomena occurring in the preceding sector of the river. The seeding of Cyanophyceae in the Charente is slow, close to the estuary, subject to tidal influences. In the Vire, this seeding occurs in the upper valley. From this factor the different composition and evolution of phytoplankton in these two coastal rivers can be understood.

In the four examples studied, zooplankton is marginal, without any truly regulatory role for phytoplankton and planktonic bacteria. It may be very different in other water courses. The zooplankton of the Seine, at Paris, is made up of rotifers and protozoa (ciliate and heterotrophic nanoflagellates), and larvae of the mollusc *Dreissena polymorpha*, the appearance of which in some seasons may be explosive. Browsing by protozoa may reach daily values equal to the net production of phytoplankton. It may also be the chief cause of mortality in planktonic bacteria. The effect of browsing was also observed in the middle course of the Loire, in 1996.

The history of eutrophication of water courses has progressed in two successive episodes. The first, from the Middle Ages to the 19th century, was the episode of canalization and reaches designed for navigation. The volume of water courses was increased in relation to the flow, at the same time increasing water transit time. The second episode, in the 20th century, was that of nutrient inputs, both nitrogen and phosphorus. Phosphorus was no longer the limiting factor of plant production. Eutrophication was accelerated by extension of transit time, followed by the input of supplementary nutrients into the aquatic ecosystems.

The interaction of regulatory and amplifying factors that vary with each stream precludes a general rule for processes of eutrophication in running waters. Each river is unique. The evaluation of eutrophication by a daily cycle of dissolved oxygen cycle or of pH is perhaps preferable to the measurements of nutrient and chlorophyll concentrations. It gives an idea of the scope of the phenomenon.

Measurements taken simultaneously on the entire longitudinal profile provide little information on the phenomenon of eutrophication. They concern different masses of water that follow from upstream to downstream, each with its own history and evolution. Extreme eutrophication, with algal blooms followed by a degeneration phase and undersaturation of oxygen in the environment, leads to a dysfunctional ecosystem and may be the cause of significant mortality. In that sense, extreme eutrophication is deferred organic pollution.

2. Trophic pollution

In the old sense of the term, *pollution* signified profanation. Anything that degrades or alters an environment pollutes it. Obviously, natural phenomena can pollute, but the term is usually reserved for degradation due to human activity.

In earlier times, pollution essentially involved the urban environment. Water pollution in France began in the 19th century with the construction of sewage systems to evacuate waste water from cities into the rivers. The 19th century was the century of water-borne epidemics: typhoid and cholera. From the end of the reign of Napoleon III, there were no fish in the Seine over a stretch of 5 km, between Saint Denis and Clichy. The construction of the Clichy main sewer led to sanitary measures in Paris but, from 1874 onward, the summer mortality of fish extended to 33 km downstream of the capital. It ultimately was aggravated further by effluents from pulp factories and starch producers extending upstream of Paris and the town settlements on the Île de France.

Pollution by organic matter results essentially from domestic and agro-food industry wastes. It can be called *trophic pollution* to the extent that it introduces nutrients into the river, just like detritus from the watershed (Chapter 11, section 1). An excess of such nutrients leads to toxic pollution (Chapter 15).

2.1. Processes of biodegradation

Organic matter, dissolved or particulate, is degraded and mineralized by decomposers, just like detritus from the watershed. Organic carbon is oxidized into carbonic anhydride, nitrogen compounds transformed into ammoniac then oxidized into nitrites and nitrates, and organic phosphorus oxidized into phosphates (Fig. 58). All these phenomena occur from upstream to downstream, over a distance that depends on transit time. Downstream, there is some self-purification, the final stages of which are complete mineralization of the organic load. These processes lead to the establishment of an equilibrium between the settlement of the water course and the ecological factors, such as existed before the organic input. Decomposers have an effect in the water column as well as on the bed.

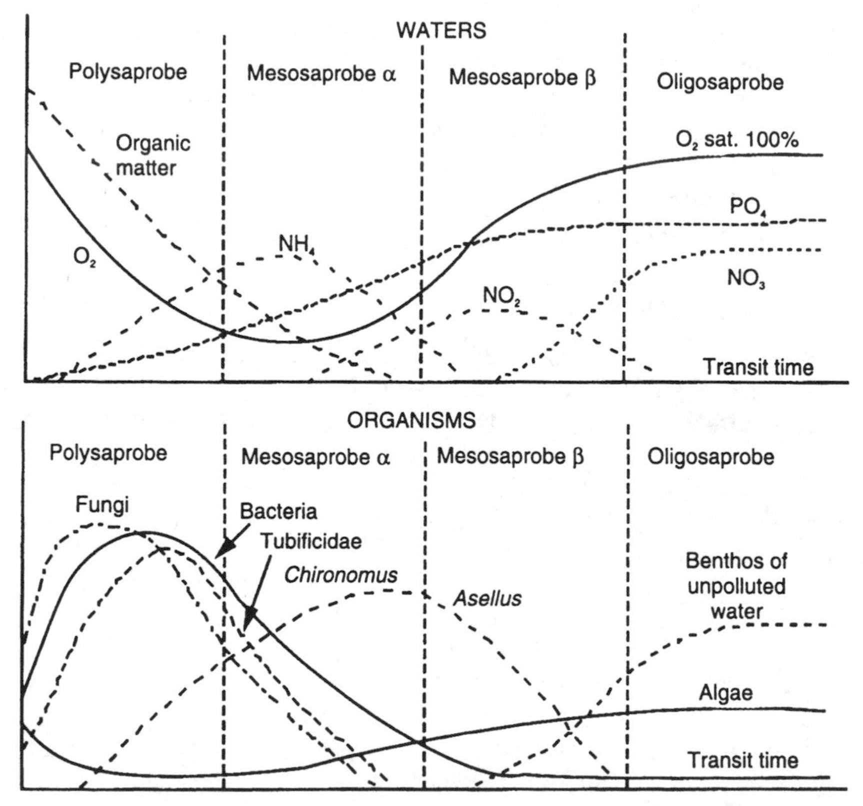

Fig. 58. Outline of the biodegradation and evolution of benthos downstream of an organic effluent (H.B.N. Hynes, 1960)

To simplify, four successive zones can be distinguished downstream of an effluent. They correspond to stages of degradation of organic matter.

Immediately downstream of the effluent, heterotrophic bacteria develop: chemo-organotrophs (greater than 10^6/ml). They are strict aerobes, such as *Pseudomonas* or *Achromobacter*, or facultative aerobes such as *Flavobacterium* or *Aeromonas*. Filamentous bacteria—*Sphaerotilus, Beggiaota*—develop in the most polluted zones. *Sphaerotilus* forms clusters fixed to the substrate that can purify 3 kg of organic matter per hectare per hour. It is in this zone, called the *polysaprobe* zone, that macromolecules are degraded. The consumption of dissolved oxygen by bacteria is high and its concentration in the water falls rapidly.

Further downstream, in the *mesosaprobe* α zone, the density of chemo-organotrophic bacteria falls, while organic compounds are oses, amino acids, and ammoniacal salts. Subsequently, bacteria such as *Nitrosomonas* appear, which draw their energy from the oxidation of the ammonia into nitrite.

Mineralization is enhanced in the *mesosaprobe* β zone with the development of autotrophic bacteria, such as nitrogen-fixing bacteria (*Nitrobacter*). Oxygen concentration approaches saturation by dissolution at the water-atmosphere interface and by algal photosynthesis.

The organic load is entirely assimilated in the *oligosaprobe* zone, which marks the return to original ecological conditions and settlements, but with an enriched mineral content (NO_3, PO_4) of the water that is available to plants, and the possibility of a eutrophication phase. The ecosystem, which was heterotrophic downstream of the organic effluent, reverts to being autotrophic.

The different saprobic zones correspond to a zonation of benthic invertebrates with, firstly, elimination of species sensitive to low oxygen concentrations, followed by a progressive reconstitution of the original settlement. In a stream in England, for example, 63 species of invertebrates were counted upstream of a polluting effluent; 9 species survived at 1.5 km downstream of the effluent, 10 at 8 km, 30 at 10 km, and 60 at 14 km.

In the polysaprobic zone, the invertebrates are detritivores, feeding on organic matter and tolerant of low oxygen concentrations: Oligochaeta Tubificidae, larvae of Diptera Eristales, and *Chironomus* of the *thumni-plumosus* group. Chironomus have a particular respiratory pigment (erythrocruorine) that can fix oxygen at very low concentrations. Although species richness is very low, the quantity of food available allows a high density of individuals. For example, 400,000 Oligochaeta and more than 100,000 larvae of Diptera Eristales per m^2 were observed.

Further downstream, in the mesosaprobic zones, when the oxygen concentration increases again, ciliate and flagellate heterotrophic protozoa appear, which consume bacteria, as well as Hirudinea, Crustacea (Isopoda *Asellus aquaticus*), larvae of *Sialis* (Megaloptera), *Tanytarsus* (Diptera), and other taxa.

Finally, the return to the original settlement corresponds to an oligosaprobic zone.

This zonation characterizes trophic pollution, with a sag curve of oxygen concentration but anoxia-free, and biodegradation of the organic load by oxidation. Such pollution is possible below a limit called the *assimilation capacity* of a river. This assimilation capacity is most often evaluated from a model simulating the equilibrium between dissolved oxygen and the atmosphere.

When the organic load exceeds the assimilation capacity and leads to a state of anoxia, anaerobic biodegradation, involving fermentation, is substituted for oxidation, with toxic compound waste (Chapter 15).

To this outline of self-purification are added peculiarities due to the simultaneous action of other ecological factors, especially current, just as for eutrophication. Biodegradation and the self-purification of water occur differently in the rhithral and the potamal.

2.2. Self-purification in the rhithral

In the rhithral, self-purification in the water column itself is negligible in relation to that which occurs in the substrate. This is due to the low volume of water in comparison to the huge area of the substrate, as well as a short transit time.

The Albenche stream, for example (watershed of Lake Bourget, in Savoy), has a flow of 50 l/s except during high water, for an average width of 2.5 m and a depth of 0.2 m. It receives waste from a pig farm in a zone corresponding to a metarhithral (order of drainage 3, slope 1.43%). Self-purification occurs in the sediments and a biological cover of bacteria develops (*Sphaerotilus* and *Beggiatoa*).

The heterogeneity of the sediments (a characteristic of the rhithral) corresponds to the heterogeneity of the bacterial film and to the deposit of the particulate organic load. The deposit starts for a current speed lower than 20 cm/s.

Self-purification involves dissolved organic matter circulating at the boundary layer and between the bacterial filaments and particles of matter trapped within the bacterial cover.

The pig farm wastes in the meta rhithral are similar to an allochthonous input from a watershed, of which it is only an amplification (Chapter 11, section 1). Physical factors, including current and flow, continue to play the dominant role they play in streams of low drainage order. They govern the granulometry of the substrate and the deposit and resuspension of the organic load. They also govern the gaseous exchanges between the water and the atmosphere, the reoxygenation of the water. A low depth and turbulence ensure the permanent saturation of dissolved oxygen on the entire polluted zone of the Albenche (with the exception of the first 100 m downstream of the waste, which is slightly undersaturated in summer).

This is a phenomenon analogous to that found in the Upper Aveyron. The conditions leading to eutrophication or pollution are combined, but the primacy of physical factors characteristic of the rhithral limits their effects. Self-purification of organic loads by bacteria in the small water courses seems minor in relation to the temporary storage and redistribution of matter.

The primacy of physical factors is found also in benthic fauna (Fig. 59). The absence of undersaturation of dissolved oxygen and of toxic compounds resulting from fermentation leads not to a total disappearance of the original community, but rather to a quantitative alteration with a reduction in species richness. The organisms that consume and filter organic matter (Chironomidae and Simuliidae) constitute the major part of the settlement in summer, at 150 m downstream of the effluent. At 1 km downstream there are also Tubificidae. Algae and Ephemeroptera reappear from the second kilometre, and Amphipoda, Plecoptera, Coleoptera, and Trichoptera from the third. A polysaprobe zone does not really exist. Species richness is greatest at the third kilometre downstream of the effluent. The organic load favours detritivores and filter feeders, without completely eliminating other species. There is a reorganization of the community rather than a total replacement.

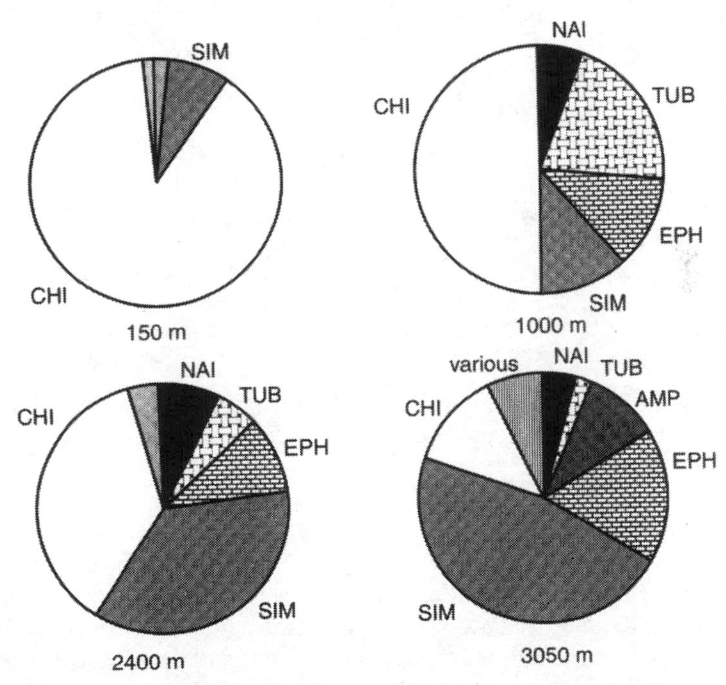

Fig. 59. Evolution of benthos on the Albenche stream, downstream of an organic effluent (D. Fontvieille, 1987). NAI, Oligochaeta Naididae. TUB, Oligochaeta Tubificidae. AMP, Amphipoda. EPH, Ephemeroptera. SIM, Simuliidae. CHI, Chironomidae.

2.3. Self-purification in the potamal

In the potamal, the depth of the water column becomes significant, while the role of the current as a factor of erosion and redistribution diminishes.

Self-purification occurs in the substrate (fixed bacteria) as well as in the water column (planktonic bacteria). In the latter, the current intervenes and determines the transit time.

The depth of the water column and the slow current limit gaseous exchanges at the water-atmosphere interface and consequently reoxygenation of the water. For the same volume of water, the assimilation capacity of the potamal is less than that of the rhithral. The oxygen concentration follows a sag curve (Fig. 60), with a deficit that may be significant during low water. The floristic and faunistic succession downstream of an effluent corresponds to that of sapobric zones.

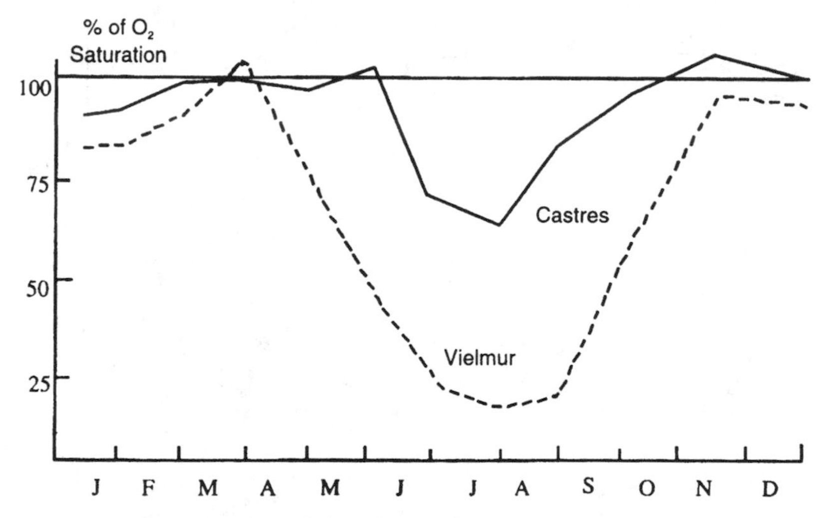

Fig. 60. Annual evolution of dissolved oxygen concentrations on the Agout (Tarn), from Castres and Vielmur, downstream of the confluence with the Thoré (Capblancq and M. Cassan, 1979)

The mechanisms and consequences of bacterial self-purification have been studied extensively in the Seine, from downstream of Paris to the estuary (Programme Piren-Seine, 1989-1992). The Achères processing plant at present deals with 75% of the domestic pollution of Paris and releases its effluents into the Seine 2 km downstream of the Maisons-Laffitte (km 708).

Self-purification of the Seine in summer

Transit time in the Seine has been measured at 40 h, in summer, from km 708 to km 847, downstream of Achères (flow of 190-200 m^3/s) (Table 20). It may reach 3 d during severely low water flow (80 m^3/s).

At 8 km downstream of the polluting effluent, 50% of the suspended matter is deposited on the bottom and is recirculated during high water.

Table 20. Biodegradation assessment of the organic load of the Seine in summer, from Achères (downstream of Paris) to the estuary, in t carbon/d. DOC, dissolved organic carbon. BDOC, biodegradable dissolved organic carbon. RDOC, resistant dissolved organic carbon. POC, particulate organic carbon (P. Servais, J. Garner, A. Barillier and G. Billen, 1993).

OISE 9 t c/d Phytoplankton 1.5 Bacteria – BDOC 1.5 RDOC 4 POC 2	SEINE 58 t c/d Phytoplankton 10 Bacteria 1 BDOC 12 RDOC 22 POC 13	ACHÈRES STATION 88 t c/d Bacteria 5 BDOC 30 RDOC 18 POC 35
	Maisons-Laffitte-Andrésy reach (km 706-721) Input 155 t c/d Output 108.5 t c/d	Biodegradation (20 t c/d) Sedimentation (28 t c/d)
% of algal C: 9.2	Phytoplankton 10 RDOC 44 Bacteria 3 POC 20 BDOC 31.5	
	Andrésy-Méricourt reach (km 721-765) Input 108.5 t c/d Algal production 28 t c/d Output 85 t c/d	Biodegradation (45 t c/d) Sedimentation (11 t c/d)
% of algal C: 21.2	Phytoplankton 18 RDOC ~42.5 Bacteria 2 POC 15 BDOC 8	
	Méricourt-Poses reach (km 765-847) Input 85 t c/d (of which algal production 28 t c/d) Output 114 t c/d	Biodegradation (30 t c/d Sedimentation (5 t c/d)
% of algal C:43.8	Phytoplankton 50 RDOC 40 Bacteria 4 POC ~10 BDOC 10	
	Estuary	

In summer, suspended matter seems to flow slowly near the bottom, because of eddies caused by river traffic.

In the sediment, the superficial zone remains aerobic. Below, in anaerobiosis, the nitrates and later the sulphates become reduced. Deeper still, organic carbon is reduced in turn, accompanied by the release of methane. At Maisons-Laffitte (km 706), 2 km upstream of Achères, nitrates and sulphates are reduced in the first 10 cm of the sediment. However, at km 749 (Porcheville), their total reduction occurs only in the first 5 cm in May, and in the first 1.25 cm in September. At 15 km downstream of Achères, phosphorus disappears from the sediments, while concentrations of carbon and nitrogen become the same as those upstream of Achères and at 40 km downstream of the effluent.

In the water column, the sudden growth of the bacterial biomass upstream of Achères is due to the input of allochthonous bacteria (larger than autochthonous bacteria) from the processing plant. They represent 75% of the bacterial biomass downstream of the effluent but regress rapidly because of sedimentation and preferential browsing by Protozoa. Protozoa are released along with the effluents, in the same way as bacteria. Allochthonous bacteria are the chief cause of the biodegradation of the organic load downstream of Achères and the deoxygenation of the water column. In the final analysis, it is the terminal phase of a self-purification state already achieved for the most part in the processing plant. Organotrophic activity is greatest in the Achères-Andrésy reach (13 km) and high in the Andrésy-Méricourt reach (44 km), while in the Méricourt-Poses reach (82 km) only the organic load resistant to biodegradation survives.

Nitrogen in the Achères effluent is present in organic and ammoniacal form. Ammonia concentration is nearly stable until km 785. Beyond that, the process of nitrification is accelerated and ammonia concentration falls, while that of nitrates rises.

Self-purification of the Seine in winter and spring

Winter and the beginning of spring are marked by the reduction of both transit time and bacterial activity due to low temperatures. In the Achères-Andrésy reach, the effluents are more diluted than in summer (because of greater flows of the Seine), but sedimentation of the organic load is lower and re-transported to the Andrésy-Méricourt reach. Self-purification is prolonged over a greater distance downstream than in the summer. In spring, 57% of the biodegradable organic load that entered the Achères-Andrésy reach still survives at Méricourt, against 18% in summer.

It must be noted that water quality in the Seine has improved considerably over that prevailing in the 1970s. In 1965, Achères processed only 30% of the urban wastes of Paris. In 2000, it processed 75%. Between Paris Notre-Dame and Mantes, oxygen concentration in the water was less than 50% of the saturation from April to November. Self-purification of wastes from towns located upstream of Paris reduced pollution in the capital. Furthermore, the construction of barrage-reservoirs with a capacity of 800×10^6 m^3 on the upper basin of the Seine led to the release of high water and increased the flow of the Seine by 50 m^3/s for 6 months of the year. The assimilation capacity of the Seine was thus improved by the dilution of effluents.

3. Eutrophication and trophic pollution: two sides of the same problem

Eutrophication and trophic pollution both correspond to an increase in the nutrient content of water, an increase in the energy potential of the ecosystem, but in two different forms. In both cases, it is the increase in water transit time that determines the phenomenon.

In eutrophication, phytoplanktonic production is made possible by mineral nutrients. When they are in excess (especially phosphorus), temperature and light amplify primary production. The ecosystem becomes autotrophic, oxygen production by photosynthesis being greater than the loss of oxygen by respiration (P/R ratio > 1). This represents the daily cycle of dissolved oxygen, with daytime oversaturation. However, extreme eutrophication may lead to phases of algal explosions, blooms, followed by degenerative phases of mortality and biodegradation of phytoplankton. Eutrophication then becomes a deferred trophic pollution.

In the phenomenon of biodegradation that follows a trophic pollution, heterotrophic bacteria—chemo-organotrophic and chemotrophic—develop. They consume oxygen, and the sag curve of oxygen concentration testifies to bacterial activity. The ecosystem is heterotrophic, but the final result of the biodegradation is an input of mineral nutrients, a source of deferred eutrophication.

By two different routes—mineral or organic nutrient inputs—the energy potential of the ecosystem is effectively increased. The two phenomena may succeed each other on the longitudinal profile of a river. They may also be simultaneous.

On the Seine, for example, from Montereau (km 634) to the entry of the estuary (km 847), the composition of phytoplankton is relatively homogeneous: Diatoms dominate in spring and Chlorophyceae in summer. The algal biomass that crosses in summer upstream of Achères represents 10 t of organic carbon per day. In the Achères-Andrésy reach, organotrophic processes largely dominate, without increase in algal biomass, with a significant oxygen deficit: the ecosystem is heterotrophic. In the Andrésy-Méricourt reach, organotrophic activity is still significant, but net algal production represents a daily input of 28 t of carbon. The oxygen produced by photosynthesis is less than its consumption (P/R < 1) but a concentration of 15 to 50% of saturation of the water column is maintained, and the fish that are less demanding in terms of the dissolved oxygen level survive (e.g., roach). The reach is a transition zone between heterotrophy and autotrophy, but with an unstable equilibrium: a rainy interlude in summer could lead to a reduction in photosynthesis and an increase in the pollutant load by leaching from urban networks.

In the Méricourt-Poses reach, the daily input of organic carbon by net primary production is 64 t/d, and the loss by respiration only 20 t (P/R ratio > 1). Each day, 18 t of algal carbon enters the reach at Méricourt, and 50 t leaves it at Poses. Eutrophication is dominant. From Achères to Poses, over 140 km, the river moves in summer from an overall heterotrophic ecosystem, with biodegradation dominant, to an overall autotrophic and eutrophic system. The latter is considerably supplied with nutrients by the mineralization of the organic load released at Achères.

In water courses without significant anthropic inputs, light, oxygen concentration, and dissolved salts play only a secondary role in the functioning of the ecosystem (Chapters 10 and 11). Physical factors such as current, substrate, and temperature are dominant, particularly in the rhithral.

When the current is no longer a limiting factor, by its erosive action or very long transit times, the role of chemical factors, especially of nutrients, becomes predominant in rivers. These factors control the energy cycle of the ecosystem: trophic pollution or eutrophication depending on their organic or mineral nature.

From upstream to downstream, nutrients (especially nitrogen and phosphorus) pass alternatively from a mineral form to an organic form across the food webs. This is expressed as the concept of *spiralling* by Etwood et al. (Chapter 11, section 4). The concept has a limited value for understanding the functioning of ecosystems regulated by physical factors, but it is useful when nutrients become the regulatory factors of primary and secondary production. The cycling time of these nutrients is then an essential factor of the overall production of living matter on the longitudinal profile of a river. This overall production is as high as the biosynthesis-biodegradation cycle is short. Etwood relates the cycling to a distance travelled by the water. It is in fact related to transit time.

15

Toxic Pollution

1. Outline and definitions

Toxic pollution originates from substances that cause alterations in the molecular and cellular functioning of organisms or in the entire organism. The action mechanisms of pollutants affecting molecular and cellular functions are the subject of ecotoxicology, a specialized discipline whose scope goes well beyond this work.

In the aquatic environment, toxins penetrate by tegumentary and respiratory routes, and by ingestion. The toxicity of a pollutant is variable.

- *Acute* toxicity leads to the quick death of the individual (e.g., carbon oxide, cyanide, parathion). Such toxicity is observed after accidental spills such as the pollution of the Rhine in 1986.
- In cases of *subacute* toxicity, all the individuals present symptoms of poisoning, but some part of the population survives.
- *Chronic* toxicity occurs over a long term and is the cumulative effect of low concentrations leading to insidious disturbances.

The toxicity of several pollutants together may be greater than the sum of toxicities of each of them (copper and zinc salts, for example). Such an effect is called *synergetic*. It may also be lower than the sum of toxicities of each pollutant. Such an effect is called *antagonistic*.

2. Toxic pollutants

An environment may be altered so that it no longer meets the vital needs of organisms. This is the case especially with a dissolved oxygen deficit, which leads to asphyxiation. Mechanical pollution (deposit of mineral or organic particles) also leads to asphyxiation through blockage of the branchiae. In an acidified environment (due to acid rain), the pH becomes incompatible with the life of organisms.

The presence or excess of certain substances has a direct consequence in the death of individuals or the regression of communities. Such pollution, usually chemical in nature, comes from two sources:

(1) Some compounds that are harmless or even necessary to organisms at concentrations natural to the river (Chapter 10) become toxic at high concentrations. While geochemical phenomena have long remained relatively constant in nature, intensive industrialization during the

20th century had the effect of releasing increasingly high quantities of salts into water bodies and courses. In the Rhine, 60,000 t of dissolved salts daily cross the Germany-Holland border. This river releases 100 t of mercury and 1100 t of arsenic into the North Sea each year.

The average level of cadmium in surface rock is about 50 μg/kg. Cadmium is at present a by-product of the zinc industry, and it is a waste effluent that is released into water courses. Superphosphates used as fertilizers contain cadmium at a rate of 0.5 to 170 μg/kg depending on their degree of purity. Cadmium concentration in fish of the Rhone varies from 0.25 to 0.70 μg/kg (as opposed to 0.05 to 0.3 μg/kg in non-polluted tributaries).

(2) In addition to these natural substances at abnormal concentrations, there are new chemical compounds produced by industries, such as detergents, pesticides, and hydrocarbons. All these substances are transported by water. When they are accidentally spilled into water, the pollution is transitory and the ecosystem is reconstituted more or less rapidly by recolonization, just as after an exceptional flood. This is what occurred in the Rhine in 1986 after the accidental spill of insecticides and fungicides at Basel, the effects of which were observed over 800 km.

Whether there are abnormal concentrations of natural substances or of new chemical compounds, organisms have not developed general adaptive strategies against pollutants. The sensitivity to each pollutant varies according to the taxa. Some methods of evaluating water pollution are based on the specific sensitivity of organisms to pollutants.

There is a historical sequence of appearance of pollutants in the water. Domestic pollution (faecal contamination) is the oldest form: in ancient Rome, the Tiber was already practically a sewer. Organic pollution from agro-food industries (e.g., dairies, canneries) followed in the 19th century, and subsequently there was pollution from dissolved salts, organic micropollutants, radioactive substances, and water acidification.

2.1. Toxic organic pollution

Trophic pollution in the potamal leads to a sag curve of dissolved oxygen concentrations (Chapter 14). When the capacity to assimilate the organic load is exceeded, anaerobic biodegradation is substituted for oxidation, with complex redox phenomena. Acids, alcohols, or aldehydes of low molecular weight are formed (e.g., lactic acid, formic acid), then methane, hydrogen sulphide, and ammonia, which cannot be oxidized in an anaerobic environment. Intermediate compounds appear during fermentation: phenol, indols, mercaptan. This phenomenon is observed in the sediments of the Seine downstream of Paris.

Toxic pollution is due to the lack of dissolved oxygen and high concentrations of compounds such as ammonia, hydrogen sulphide, and phenols. Species most sensitive to oxygen deficit are eliminated, while those that are tolerant of or capable of fixing oxygen at low concentrations increase in number (Oligochaeta Tubificidae, Chironomidae of the *thumniplumosus* group).

Mortality due to lack of oxygen first affects species of the rhithral. In rainbow trout, the minimal concentration of O_2 needed for survival of 84 h is 3 mg/l at 16°C and 1.9 mg/l at 10°C. Fish of the potamal are more tolerant:

— perch (*Perca fluviatilis*) survive with 1.35 mg/l at 16°C, 1 mg/l at 10°C;
— carp (*Cyprinus carpio*) survive with 1.75 and 0.5 mg/l respectively;
— tench (*Tinca tinca*), survive with 0.55 and 0.35 mg/l respectively.

An oxygen concentration of 7 mg/l is considered necessary for salmonids, while 3 mg/l suffices for cyprinids.

Ammonia and hydrogen sulphide are more toxic in non-ionized form (NH_3 and SH_2) than in ionized form. The following equilibrium reaction occurs:

$$NH_3 + H_2O \leftrightarrow NH_3OH \leftrightarrow NH_4 + OH$$

Equilibrium moves towards NH_3 when the pH increases. For hydrogen sulphide, it moves towards SH_2 when the pH falls. The sensitivity of species to toxins from fermentation is variable. Fish are more sensitive to ammonia than invertebrates (Table 21).

Table 21. Sensitivity of some fish and invertebrates to ammonia at 14°C and pH 7.5 (H. Stammer and K. Wuhrmann, 1958)

	LC_{50} in 16 h 30 min
Salmo trutta	0.95 mg/l
Oncorhynchus mykiss	0.6 mg/l
Perca fluviatilis	1 mg/l
Gammarus pulex	7 mg/l
Perla cephalotes	14 mg/l

Hydrogen sulphide appears only in anaerobiosis. In running water, it persists only in the sediments. In the presence of ferrous iron, there is formation of ferrous sulphide, which gives sediments a blackish colour. Some Diatoms, Cyanophyceae, Flagellata, and Ciliata tolerate low concentrations of SH_2. However, at just a few mg/l, only bacteria survive.

At a phenol concentration of 10 mg/l, only bacteria survive, and even they disappear in 72 h at a concentration of 20 mg/l.

With ammonia, hydrogen sulphide, or phenols, the maximum concentrations allowing an indefinite survival are obviously much lower than

LC_{50} (lethal concentrations for 50% of individuals in a given time period), determined experimentally. The concentration limit for indefinite survival of Salmonidae is 0.1 mg/l of non-ionized ammonia and 0.5 mg/l of ionized ammonia.

2.2. Saline pollution

Saline pollution is most often caused by chlorides. Chlorides occur in significant quantities in domestic wastes, but especially in industrial wastes, where chlorine is associated with sodium, potassium, or calcium. In France, this type of pollution occurs notably in the Moselle river, reaching an international dimension in the Rhine (540,000 t of chlorides released per year in Alsace).

In natural waters, dissolved salts appear rather as factors of plant production and eutrophication (e.g., phosphates, nitrates, calcium salts). At high concentrations, they affect osmotic pressure of organisms and they are toxic.

During osmotic pressure, the effect of dissolved salts (particularly of chlorides) is analogous to that observed in brackish waters: replacement of stenohaline species by euryhaline species. The production of a macrophyte such as *Elodea canadensis* falls significantly at a chlorine concentration of 1 g/l, while Amphipoda *Gammarus pulex* is eliminated. In a German stream polluted by chlorides, euryhaline Oligochaeta are dominant. In an arid or semi-arid region, intense leaching of soils also gives rise to an increase in the salinity of water.

2.3. Chemical pollution

Many organic compounds created by the chemical industry do not exist in the natural state, and their toxic effect is not predictable.

Detergents are made up of surfactants (most often anionic, e.g., alkylsulphates, alkylsulphonates) whose action is complemented by additives such as polyphosphates, carbonates, and silicates. Polyphosphates contribute to the eutrophication of waters. Surfactants limit the dissolution of oxygen to the water-atmosphere interface. They have low toxicity: 10 mg/l are required to inhibit bacterial flora, and 50 mg/l to inhibit phytoplankton. In fish, a decline in surface tension of the medium modifies the respiratory exchanges at the gills. It also leads to the disappearance of insects living on the water surface (e.g., Hemiptera *Gerris*).

Pesticides—fungicides and insecticides—are much more toxic than detergents, since their purpose is by definition the destruction of crop pests and since they are carried into the water by runoff.

At a concentration of 25 µg/l, organophosphate insecticides are considered non-toxic for Oligochaeta, Mollusca, and Crustacea Copepoda and Ostracoda. Culicidae and Cladocera larvae, however, are eliminated at this concentration, as are Amphipoda Gammaridae (LC_{50} 9 µg/l at 96 h). Derivatives of phenoxyacetic acid (2,4D) are used as herbicides. They are toxic at concentrations of 0.5 to 4 mg/l depending on the species (LC_{50} 0.5 mg/l in 24 h for fish and *Gammarus pulex*).

Possible toxic effects occur in the short term when biodegradation of the pesticide is rapid (50% of organophosphates are degraded in the soil within 48 h). They occur over the long term when micropollutants degrade slowly and accumulate in the tissues of organisms.

2.4. Cumulative effect micropollutants

When the excretion rate of a poorly biodegradable or non-biodegradable pollutant is lower than the rate of chronic contamination, the pollutant accumulates in the organism.

Cumulative effect micropollutants include certain pesticides (organochlorine), anions (fluoride), and heavy metals. Their toxic effect is no longer directly linked to their environmental concentration, but dependent on their accumulation in the organism, which increases with time till it reaches lethal levels. It is also dependent on transfer along the food webs through the predator-prey relationship. Short-term toxicity tests (LC_{50}) cease to be valid when the toxic effect is due to the non-degradation of pollutants and its bioaccumulation in the tissues.

Organochlorine pesticides degrade slowly (50% of DDT persists in water after 10 years). Before such pesticides were banned, their concentrations in the water were low (50 µg/l of lindane and 280 µg/l of dieldrin on average in Britain, lower than short-term toxic concentrations). However, for an environmental concentration of 14 µg/l, some 5 mg/kg were measured in algae (concentration factor or FC 350), 7 to 8 mg/kg in planktonophagous fish (FC 500 to 650), and 22 to 220 mg/kg in carnivorous fish (FC 1500 to 15,000).

Toxic heavy metals are essentially mercury, copper, lead, cadmium, zinc, cobalt, and chromium, in the form of salts. They tend to fix themselves to organic sediments and clays and are released during high water.

The most toxic heavy metals are those that can form stable organometallic complexes, which inhibit enzymatic activity. For example, mercury chloride ($HgCl_2$) incorporated into the sediments is transformed under bacterial action into methylmercury (CH_3HgCl), which has a much higher capacity for direct penetration.

Many simple experimental models have enabled the study of transfers of cumulative effect micropollutants within food chains. For example:

- *Chlorella vulgaris* (alga) to *Daphnia magna* (filter-feeding Cladocera) to *Gambusia affinis* (planktonophagous fish) to *Oncorhynchus mykiss* (rainbow trout, a carnivorous fish); and
- *Ranunculus aquatilis* (macrophyte) to *Lymnaea peregra* (Mollusca Gasteropoda) to *Orconectes limosus* (crayfish).

Algae are rapidly contaminated by direct penetration of micropollutants, given their large area of contact with the water in relation to their volume. At a density of 2×10^6 cells/ml and a temperature of 18°C, *Chlorella vulgaris* absorbs all the methylmercury present at a concentration of 1 μg/l in 24 h. The FC in algae may exceed 100,000. It reaches 15,000 in 10 d in *Elodea canadensis* leaves. The bioaccumulation capacity of Bryophytes, which are perennial and slow-growing, makes them useful in detecting traces of heavy metals in water.

For the consumers, contamination by the trophic route is added to direct contamination and becomes dominant. In the *Chlorella-Daphnia-Gambusia-Oncorhynchus* food chain, the FC in the consumer of the third order is around 10,000 in 30 d and at 18°C, for a methylmercury concentration in the water of 1 μg/l.

Direct contamination is dependent on certain environmental factors or on the defence reactions of organisms. The absorption of lead salts by Gasteropoda *Ancylus fluviatilis* is greater in waters with a granite substrate than in waters with a chalky substratum. The cell wall is less permeable in a chalky medium, because the active absorption of ions is not necessary. Defence reactions also exist (e.g., formation of mucus), limiting the penetration of pollutants.

In fish, direct absorption of metallic salts through the skin and gills is attributed to active transports when trace amounts of the salts are present in the water, or even to simple passive penetration when the concentration is higher. Beyond a certain threshold of concentration, defence reactions appear that tend to reduce the permeability and passage of metals through the cell membranes. This explains, for example, the contamination of fish in the Lot by heavy metals (copper, zinc, lead, and cadmium) carried by a tributary, the Riou-Mort. The concentration of metal salts decreases over 200 km downstream of the Lot-Riou-Mort confluence, while their accumulation in the tissues of fish is higher downstream than upstream.

The trapping of metallic salts in sediments has a significant impact on trophic contamination. Contamination is higher in burrowing fish than in omnivores and carnivores (Table 22).

Table 22. Average levels of heavy metals in fishes of the Lot downstream of the Riou-Mort, as a function of food regime (R. Labat et al., 1977)

	Average level in tissues (mg/g)		
	Carnivores	Omnivores	Burrowing species
Copper	12.3	12.2	23
Zinc	66.6	144.3	122.8
Lead	0.88	1.05	3
Cadmium	5.5	33.6	45.1

2.5. Acidification of waters

The acidification of rivers and lakes was first observed in Scandinavia. In France, the Vosges is presently the region most affected. Acid rains, responsible for the acidification of waters, are the consequence of the use of fossil fuels, carbons, and fuels in which the sulphur level reaches 5% for certain carbons, and at least 3% for certain heavy fuels. Their combustion releases sulphurous anhydride (SO_2) and nitrogen oxides (NO and NO_2) into the atmosphere. These are transformed into sulphuric and nitric acids and are carried to the soil and water by rain and snow. Thus, a pH of 3.8 to 4.8 has been measured in rain water, and 4.4 to 4.9 in the water from snow melt in Canada.

On chalky substrates, the buffering power of bicarbonates reduces or eliminates the effect of acid rain. The same does not hold for crystalline substrates, where acid inputs cannot be neutralized, especially during snow melt. Acidification of waters may be permanent or temporary (depending on precipitation or snow melt).

The concentration of H^+ ions in water is expressed by pH values lower than those found normally in natural waters. pH thus becomes a limiting factor. In two highly acidified torrents of the Vosges, the average pH is 4.89 (4.22 to 5.77) and 4.78 (3.62 to 5.42). Concentrations of aluminium are 293 (100 to 600) and 354 µg/l (126 to 1060).

Studies on Ephemeroptera have shown a very high sensitivity of young larval stages and eggs to acidification. There is also a decline in fertility, growth rate, and emergence rate, together with perturbations in ionic regulation.

Leaching of crystalline soils by acid rain leads to dissolution of aluminium, iron, manganese, and copper. Aluminium and copper at high concentrations are toxic, for invertebrates as well as for protozoa and algae. Salmonids are particularly sensitive to aluminium. Concentrations higher than 1 mg/l have been observed in a torrent in the Vosges, while the reference value for drinking water is fixed at 50 µg/l (200 at most) by a European Union directive.

Ephemeroptera are the most sensitive invertebrates to acid waters (Fig. 61), unlike Plecoptera (particularly Nemouridae, Leuctridae, and Capniidae). Species diversity of Trichoptera, Coleoptera, and Diptera declines. In four torrents in the Vosges in which acidification is increasing, the number of taxa of Trichoptera falls from 14 in the least acidified torrent (Foulot) to 3 in the most acidified torrent (Grand-Rupt). That of Coleoptera falls from 8 to 2 and that of Diptera from 12 to 5. Salmonids survive only in the Foulot torrent. Conversely, Oligochaetes have low sensitivity to acidification.

From the Foulot torrent to the Grand-Rupt torrent, the total number of taxa falls from 52 to 17, and macro-invertebrate density from 933 to 295 individuals/m^2, i.e., a reduction of two thirds.

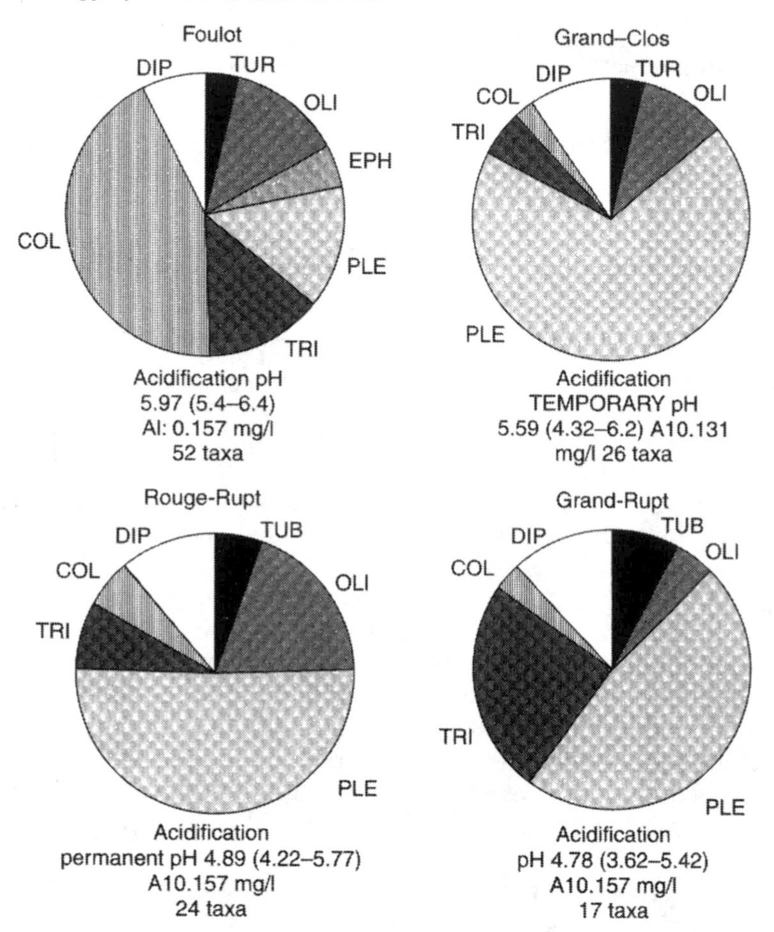

Fig. 61. Structure of invertebrate community in four torrents of the Vosges, as a function of acidification of water (S. Guerold and J.C. Pihan, 1989). TUR, Turbellaria. OLI, Oligochaeta. EPH, Ephemeroptera. PLE, Plecoptera. TRI, Trichoptera. COL, Coleoptera. DIP, Diptera.

3. Multiple pollution

A single pollution source is a relatively rare phenomenon. Industries are in fact concentrated in or near towns, and often many industries are located along a river.

3.1. The Riou-Mort[1]

The Riou-Mort, in the Aveyron, is a good example of a small river with multiple pollution. The Riou-Mort and its tributaries drain the industrial

[1] This section is based on research conducted between 1978 and 1984. Since then, some industrial sites have become inactive, and domestic and industrial wastes are now subject to processing.

basin and coal field of Decazeville, before emptying into the Lot after a journey of 23 km. At its confluence with the Lot, the flow is less than 1 m^3/s in summer, and about 90% is domestic and industrial wastewater. In winter, the average flow is around 2 m^3/s, with high water that may reach 20 m^3/s.

On its upper course, the Riou-Mort is only a stream with a substrate of pebbles and sand, crossing agricultural terrain. The settlement is rich and diverse: 26 species of benthic algae (of which 17 are Diatoms) are abundant, as well as 85 taxa of invertebrates (Fig. 62A). Among these, 8 taxa are Ephemeroptera, 8 are Plecoptera, 11 are Coleoptera, 3 are Mollusca, and 32 are Oligochaeta. Oligochaeta represent a density of two thirds of the invertebrates present, Naididae being dominant (41.7% of individuals).

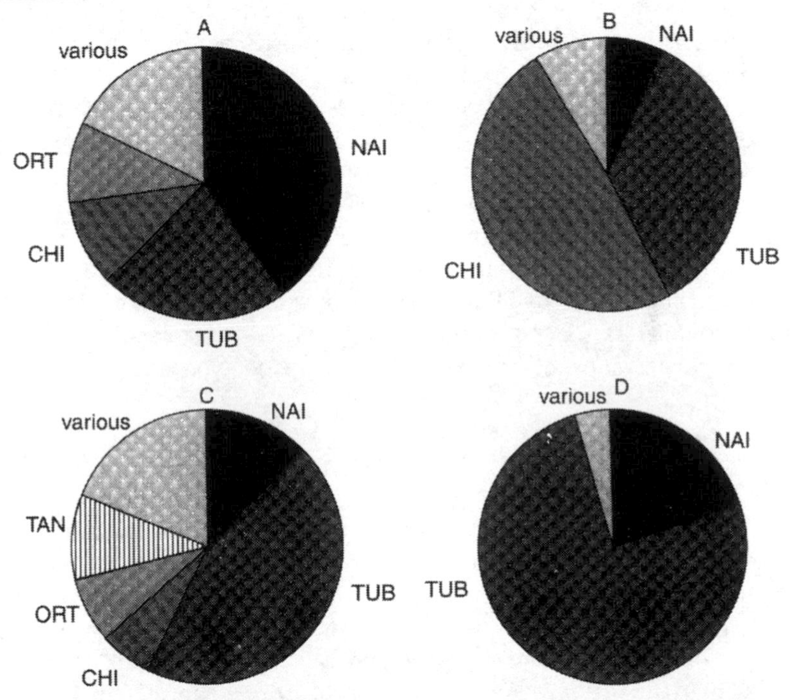

Fig. 62. Structure of invertebrate community of the Riou-Mort. A, upper course. B, km 10. C, km 15. D, km 17.5 (N. Giani, 1983, 1984). NAI, Oligochaeta Naididae. TUB, Oligochaeta Tubificidae. CHI, Diptera Chironomidae. ORT, Diptera Orthocladiinae. TAN, Diptera Tanypodinae.

At km 10, the Riou-Mort enters the Decazeville basin and collects the sewage of Firmi (pop. 2700). The substrate is muddy and organic pollution is indicated by the appearance of self-purifying bacteria such as *Sphaerotilus natans* and mesosaprobic and polysaprobic algae. From December to June, the community is diverse, with Ephemeroptera, Coleoptera, and Diptera Simuliidae present. However, in summer, the association of *Chironomus riparius* and Tubificidae represents 84% of individuals, while Naididae regress (6%). Species richness diminishes (54 taxa of invertebrates), but the density of individuals increases (Fig. 62B).

At km 15, at the entrance to Decazeville, the start of a self-purifying flow (over 5 km) leads to an increase in species richness (75 taxa), with the presence of *Helobdella stagnalis* (Hirudinea), *Asellus aquaticus* (Isopoda), and *Sialus lutaria* (Megaloptera) in summer. Simuliidae and Ephemeroptera are present in great numbers in winter. Diptera Orthocladiinae, Chironomidae, and Psychodidae are also abundant, while *Chironomus riparius* regresses. Naididae represent 17.4% of individuals (Fig. 62C). From a polysaprobe zone in summer at km 10, the river reverts to a mesosaprobe zone at km 15.

Downstream of Decazeville (pop. 7750), at km 17.5 (Fig. 62D), the Riou-Mort receives domestic wastes from the city as well as those of collieries and two steelworks (warm wastewater that increases the temperature by 7 to 8°C in summer, in relation to the upstream temperature). The river becomes rich in sulphates (600 mg/l of SO_4 on average), manganese (2.2 mg/l), iron (4.1 mg/l), and zinc (0.56 mg/l). The water turn's brownish because of particles in suspension. These seal up the interstices of the substrate, while iron precipitates. *Sphaerotilus natans* is abundant, but only four species of algae remain. The density of invertebrates is the same as upstream of Decazeville, but the community is different (35 taxa, of which 14 are represented by 10 individuals at most, and 95% of individuals are Tubificidae). There is no longer a temporary winter recolonization. It has become a polysaprobe zone, even antisaprobe, because of acid and metallic wastes.

At Viviez (km 19), there is also domestic pollution from a population of 1650. Up to the confluence with the Lot, the water remains highly coloured and the substrate is covered with thick slime. Oligochaeta are practically the only invertebrates, constituting 99.6% of individuals, of which 98.6% are Tubificidae.

Downstream of Viviez, the Riou-Mort receives a tributary, the Riou-Vieux. It transports, in addition to the domestic waste of a population of 8000, the effluents of a chemical plant (paints) and a zinc foundry. Despite its partial neutralization, the sulphuric acid lowers the pH of the Riou-Mort from 7.5 to 6.9, while at km 19 the average concentrations of metals increase: zinc from 0.56 to 34 mg/l, cadmium from 0.02 to 0.85 mg/l, and copper from 0.05 to 0.16 mg/l. In fact, these concentrations are subject to sharp variations depending on the effluents. A single alga survives, *Hormidium rivulare* (Chlorophyceae). The settlement is random: 15 taxa in 38 individuals. This is an antisaprobe, azoic zone.

Before its confluence with the Lot, the Riou-Mort receives waters from a power station extracted from the Lot, and its pollution is diluted. The beginnings of recolonization appear, fluctuating over time, probably from the Lot.

3.2. Conclusions

Two types of pollution are seen on the Riou-Mort: organic and chemical. The organic pollution, which is domestic, increases from upstream to downstream. It is linked to the riverside population, which is 2700 at km 10, 10,450 at km 17.5, 12,100 at km 19, and 20,100 at km 20.

Upstream of Firmi, taxa at low or moderate numbers are predominant, indicating a diverse community. At Firmi, the 10 species of Coleoptera disappear, while Crustacea (Copepoda and Ostracoda), Naididae, Ephemeroptera, and Plecoptera regress. Their survival results from the winter and spring recolonization. The Riou-Mort is at this point a polysaprobe environment, with a *Chironomus riparius* and Tubificidae association in summer.

Upstream of Decazeville, a partial self-purification restores the Riou-Mort to the mesosaprobe state, with an increase in taxonomic richness.

Downstream of Decazeville, the pollution is both organic and chemical (SO_4 and heavy metals), without a winter and spring recolonization. Crustacea, Ephemeroptera, and Plecoptera disappear, while Naididae and Diptera remain at a minute percentage. The environment again becomes polysaprobic. Tubificidae, practically alone, are resistant to organic as well as limited chemical pollution. Downstream of the confluence with the Riou-Vieux, high SO_4 and heavy metal concentrations and their abrupt fluctuations as a function of effluents render the Riou-Mort azoic.

The growing proportion of Tubificidae from upstream to downstream indicates an increase in organic pollution and reflects their resistance to pollution. The resistance, however, depends on the species. Out of 14 species of the Riou-Mort, 10 colonize the environment upstream of Firmi, 7 downstream of Firmi, 10 upstream of Decazeville, 7 again downstream of Decazeville, and only 6 at Viviez.

From Firmi, three species of Tubificidae disappear and one, *Tubifex ignotus*, regresses abruptly. On the other hand, *Limnodrilus profundicola* is favoured by a pollution that can be considered moderate and subsequently regresses downstream. *Tubifex tubifex* progresses from Firmi and subsequently survives at the same level.

The percentage of *Limnodrilus hoffmeisteri*, *L. claparedeanus*, and *L. udekemianus* increases from upstream to downstream. These three species and *T. tubifex* appear to be the most resistant to chemical as well as organic pollution.

If the evolution of communities in the acidified waters of torrents of the Vosges is compared with that of communities in the Riou-Mort, a reduction in the species richness of Coleoptera is observed at first associated with water acidification, but Coleoptera totally disappear with organic pollution (the slimy nature of the substrate from Firmi is perhaps partly responsible for this disappearance, characteristic of species of the rhithral).

Ephemeroptera are more sensitive to acidification than Plecoptera, but both groups are highly sensitive to organic pollution. The species richness of Trichoptera and Diptera declines with acidification. That of the Diptera also declines on the Riou-Mort, while species richness of the Trichoptera disappears.

Oligochaeta seem to be the least sensitive to pollution, to acidification as well as organic and chemical pollution. Tubificidae are the last invertebrates to survive in the Riou-Mort before inputs from the Riou-Vieux.

The varying sensitivity of species to pollution and the species richness of communities constitute the foundation of biological methods of evaluating pollution.

16

Biological Methods of Evaluating Pollution

Water quality is one of the components of the environment, along with the quality of sediments and microhabitats. The term *water quality* is ambiguous. It depends in fact on the use to which the water is put (drinking, leisure, fishing, irrigation) and is often confused with environmental quality (which takes the biota into account).

The aim of biological methods of evaluating water quality is to detect any physical or chemical degradation of the environment by studying the composition of the settlement, its species richness, and the appearance or disappearance of species. Some methods of evaluation use biochemistry or ecotoxicology. The methods used most widely, however, are biocoenotic methods.

1. Methods using biochemistry or ecotoxicology

Some methods are designed to evaluate the trophic state of an environment. The autotrophy index, for example, indicates the proportion between autotrophic and heterotrophic components of an environment, in relation to biomass and chlorophyll concentration.

The BOD_5 (biological oxygen demand in 5 d at 20°C, expressed in mg/l) corresponds to the oxygen consumption needed for oxidation of organic compounds that are easily biodegraded (e.g., glucides) by self-purifying bacteria.

Toxicity tests are used to classify pollutants according to their degree of toxicity, while bioassays are designed to survey polluted environments. Bioassays consist of evaluation of chronic toxicity of a polluted environment on an organism and association of the effect on the organism to the cause (i.e., pollution). This is possible if, through toxicity tests, the relationship between the pollutant concentration and the mortality rate of individuals in a given time (LC_{50}) is known.

The organisms used to evaluate pollution are fish (measurement of activity by respiratory rhythm or swimming endurance), bacteria, algae and protozoa (measurement of growth rate), Cladocera (Daphnia test), and Batrachians (detection of genetic mutations).

Some bioaccumulator organisms assimilate pollutants that exist in trace levels that cannot be detected chemically (pesticides, heavy metals). Fish, Mollusca, Oligochaeta, and slow-growing macrophytes (especially Bryophytes) are thus used as micropollutant tracers.

2. Biocoenotic methods

Biocoenotic methods are based on the preponderant role of ecological factors in the dynamics of communities. Communities are considered the synthetic expression of various environmental factors, and their structure consequently reflects the physicochemical characteristics of each environment.

Two phenomena are manifested jointly downstream of a polluting effluent:

— the development of populations showing an affinity for the compounds introduced (organic matter, sulphurous and ferruginous compounds, or minerals); and

— the regression or disappearance of species variously sensitive to physicochemical modifications of the water and the substrate.

From the growth of certain populations and the regression or disappearance of others, lists of indicator species can be drawn up for the level of pollution, based on their degree of sensitivity to pollution. These lists most often involve a part of the community consisting of species that are phyletically related.

Another approach taking into account the entire community, the biocoenosis, is simultaneously qualitative by way of its indicator species and quantitative by way of its species richness.

2.1. Comparative analysis of communities

Pollution can be evaluated by comparison of the community with that of a natural environment or a typology of the natural environment.

Communities of acidified torrents of the Vosges, for example, have been compared to those of the Foulot (also in the Vosges). The evolution of the perturbation may be followed in space (from one water course to another or on a longitudinal profile) and in time.

Analyses that refer to a typology assume that it corresponds to natural communities. In Europe, the natural typology may be that of Huet for fish (slope-width rule, Chapter 8), Illes and Botosaneanu's zonation, or even Verneaux's typology (Chapter 12). From comparisons of altered csommunities in relation to a reference typology, a new typology can ultimately be established as a function of the eutrophication or pollution level. This can be done from various indexes.

Communities can be compared two by two using similarity indexes. There are about 15 indexes, but the simplest, and probably the most efficient, is the Jaccard index (Table 23):

$$IS = \frac{N_c \times 100}{(N_a + N_b)}$$

where N_a is the number of species in community a, N_b is the number of species in community b, and N_c is the number of species common to a and b.

Table 23. Comparative analysis of benthos of the Riou-Mort stream by the Jaccard similarity index (N. Giani, 1983, 1984)

	Downstream of Firmi	Upstream of Decazeville	Downstream of Decazeville	Downstream of Viviez	Upstream of the Lot
Upstream of Firmi	33.1	35.8	18.9	23.5	19.4
Downstream of Firmi		35.1	30.3	27.4	24.7
Upstream of Decazeville			32.3	26.8	20.6
Downstream of Decazeville				36.8	24.8
Downstream of Viviez					27.4

Diversity indexes (Fisher, Shannon, Simpson, etc.) are used to classify communities as a function of their species richness. The models of Motomura (1932), MacArthur (1957), and Frontier (1976) are range-frequency diagrams. Frontier's model (logarithm of relative frequency and of the range of each species) seems to be the one that can best be adapted to all types of communities.

2.2. Methods based on the vicariance of species belonging to a single group

Diagnosis by indicator species is based on their physiology and ethology. A biological indicator corresponds to a population or set of populations indicating the state of an environment by their qualitative and/or quantitative characteristics. From variations in these characteristics, possible environmental disturbances can be detected. The recognition of biological indicators calls on methods peculiar to ecology: demography, species richness, and abundance.

Methods of evaluation by species or biological indicators are based on the succession of algae, macrophytes, invertebrates, or fish downstream of a polluting effluent. This succession indicates the gradual self-purification of the environment and its level of saprobity (Chapter 14). However, in a community, one species rarely represents a single saprobic level. This is why most authors attribute an indicator weight, or indicator power, to each species depending on the number of saprobic levels it occupies. By this means they distinguish true indicator species from those that are more ubiquitous.

The sensitivity of plants to chemical factors of the water and sediments makes them useful as biological indicators, just as bacteria indicate levels of pollution (or eutrophication). Diatoms are the most often used algae. They are highly sensitive to pollutants and especially to nitrogen and phosphorus compounds, but less sensitive to factors other than pollution such as the nature of the substrate. Their biological cycle is short and assimilates short- and medium-term pollution.

From species or associations of species, *pollution indexes* can be established. The diatomic indexes of Sladecek (1986) and Leclerq and Marquet (1987) use the saprobe system of Kolkwitz and Marsson (1902-1909): five levels of water, from the purest to the most polluted (Table 24). Descy and Coste's grid (1988) was established from 208 species of Diatoms, comprising 8 groups of 7 species and 4 sub-groups, and was subsequently used to set up a pollution scale from 1 to 10.

Other indexes use macroalgae (Dell'uomo, 1991, in Italy) or even algae and macrophytes together (Harding's plant score, 1987, in Britain)

Table 24. Characteristics of five saprobic levels in running waters (P. Ghetti and G. Bonazzi, 1981)

Polysaprobe	High organic pollution. Massive growth of chemo-organotrophic bacteria (10^6/ml). Low O_2 concentrations. Presence of ammonia and hydrogen sulphide. Few species. Dominance of detritivorous invertebrates (Chironomidae, Oligochaeta Tubificidae). Near absence of primary producers ($P/R < 1$).
α-mesosaprobe	High organic pollution. 10^3 to 10^6/ml chemo-organotrophic bacteria. Organic compounds: osides, amino acids, ammoniacal salts. Appearance of bacteria drawing energy from oxidation of ammonia and nitrites: a majority are Chironomidae and Tubificidae. More numerous species than in the polysaprobe zone.
β-mesosaprobe	Low organic load. Chemo-organotrophic bacteria regress in favour of autotrophic (nitrogen-fixing) bacteria, as well as detritivorous invertebrates. Increase in species richness. O_2 concentration close to saturation.
Oligosaprobe	Near total absence of organic load. Predominance of primary producers and non-detritivorous consumers. High species richness. Return to saturation of O_2 concentration ($P/R > 1$).
Xenosaprobe	Total absence of organic load.

or macrophytes on their own, as with Haslam and Wolseley's index (1981, Britain), Müller's eutrophication sequences (1990, low sandstone Vosges), Carbiener's index (1981-1983, the Alsace Ried water course), Empain's bryophyte index (1978, Belgium), and J. Haury et al.'s macrophyte indexes (1996).

The macrophyte cycle, which is longer than that of the Diatoms, integrates pollution over a period of several months. The same applies to macro-invertebrates. Indexes based on macro-invertebrates essentially use Oligochaeta and Diptera Chironomidae, which, by their mode of life, reflect water quality as well as the quality of sediments.

Oligochaeta cannot be used to detect mild pollution (such as the kind that causes the disappearance of Ephemeroptera, Plecoptera, and Coleoptera). Indexes based on Oligochaeta are simplified and used most often to evaluate the proportion of Tubificidae (the most resistant of the Oligochaeta) in communities of shifting beds: ratios between Tubificidae and other invertebrates, or even between Tubificidae and other Oligochaeta.

Chironomidae are characterized by a large number of species with highly variable sensitivities to pollution. Relatively precise pollution scales can consequently be established. Nymphal exuviae (collected in drift nets) are easier to identify than larvae and are consequently used as indicators.

Bazerque, Laville and Brouquet's chironomid index (1989) uses 24 indicator species distributed in 5 groups and also takes the current into account. Three groups are used to evaluate average to high pollution; the other two, more pollution-sensitive, are used to evaluate average and good quality environments. The indexes are calculated from a double-entry grid: the indicator groups are listed vertically, while species richness and Shannon's index are listed horizontally (Tables 25 and 26). This chironomid index has been used in the basins of Picardy, the Loire, and the Garonne, as well as the Tenzi wadi (Morocco).

2.3. Methods based on a combination of benthic macro-invertebrates

Some methods of evaluating pollution are based on all organisms combined—bacteria and fish—but most often only macro-invertebrates (> 0.5 mm) are used. The taxonomy, ecology, and degree of pollution sensitivity of most of them are quite well known. These methods are derived from Kolkwitz and Marsson's saprobe system, but with more precise pollution indexes. Most of these indexes originate from the Trent biotic index perfected by Woodiwiss (1964) for the basin of the river Trent in Britain, and the Extendens biotic index by the same author (1978). While the original methods based on the saprobe system use the affinity of organisms for organic matter only, all types of pollution are taken into account by biotic indexes. Their level of identification of organisms differs according to the groups (taxonomic units) and the difficulty of their

Table 25. Table of determination of the chironomid index (Bazerque, Laville and Brouquet, 1989)

Shannon index	0-1	1-2	2-3	3-4	> 4	
Species richness (species > 1 ind.)	1-10	11-20	21-30	31-40	> 40	
Dominant indicator species (> 10%)		Chironomid index				Pollution
Highly resistant to pollution						High to severe
A	1	2	3	4	5	High
B	2	3	4	5	6	Severe in lentic zone
C	3	4	5	6	7	Severe in lotic zone
Moderately resistant to pollution						Moderate to slight or nil
D	5	6	7	8	9	Potamal-rhithral
Slightly resistant to pollution						Rhithral
E	6	7	8	9	10	

Table 26. Indicator taxa used in the chironomid index. A, taxa intolerant of pollution. B, relatively intolerant taxa. C, relatively tolerant taxa. D, taxa resistant to pollution (Bazerque, Brouquet and Laville, 1989).

Upper part: indexes 1 to 7	
1 Most pollution-resistant species	
D *Chironomus riparius*	
2 Pollution-resistant species in lentic facies (5 *Chironomini genuini*)	
D *Chironomus annularius*	D *Dicrotendipes notatus*
D *Chironomus bernensis*	C *Glyptotendipes (Phytoten.), pallens*
3 Pollution-resistant species in lotic facies (4 Orthocladiinae + 3 *Tanytarsini*)	
D *Cricotopus bicinctus*	B *Rheocricotopus fuscipes*
D *Microspectra atrofasciata*	B *Rheotanytarsus photophilus*
C *Paratrichocladius rufiventris*	B *Rheotanytarsus rhenanus*
C *Eukiefferiella claripennis*	
Lower part: indexes 5 to 10	
4 Moderately pollution-resistant species of the potamal and rhithral (5 Orthocladiinae + 1 *Chironomini genuini*)	
D *Nanocladius bicolor*	C *Cricotopus* sp. (> 20%)
D *Cricotopus sylvestris*	C *Parachironomus arcuatus*
C *Synorthocladius semivirens*	B *Rheocricotopus chalybeatus*
5 Species less pollution-resistant or taxa practically of the rhithral (5 Orthocladiinae)	
A *Eukiefferiella* (relative abundance or number of species > 4)	B *Orthocladius (Euorthocl.) ashei*
	B *Orthocladius* (s.str.) *frigidus*
A *Orthocladius (Euorthocl.) rivicola*	B *Orthocladius (Euorthocl.) rubicundus*

identification, making the method accessible to non-specialists. The biotic index is based on key taxa (biological indicators) and the number of taxonomic units (species richness).

Several biotic indexes derive from the Trent biotic index (in Denmark, Germany, Belgium, France, Spain, and Portugal), with adaptations necessitated by the geographical distribution of organisms, the types of water courses, and the climate. In France, Tuffery and Verneaux's biotic index (1967) was inspired directly by the Trent index. The few modifications made to the classification of taxonomic units were designed to better link the method to the ecology of benthic macro-invertebrates. The samples were taken in both lotic and lentic environments, when rapid and slow currents existed in a single location. A certain number of environmental parameters were also taken into account (width of the bed, Illies and Botosaneanu's zonation) so as to verify the validity of the results. The degrees of pollution—biotic indexes—are classified from 0 (most polluted waters) to 10 (water of excellent quality). On a double entry table, 7 faunal bioindicator groups of the crenon, rhithron, and potamon are listed vertically, in order of increasing pollution resistance, while the number of taxonomic units present is listed horizontally (Table 27). The biotic index is deduced by means of the bioindicator groups as well as from taxonomic richness (Table 28).

Tuffery and Verneaux's biotic index is straightforward and easy to use. In order to make it more effective still, the following indexes are added: a global index of biological quality (GIBQ) for a shallower environment and a potential index of biological quality (PIBQ) for a deep water environment (using artificial substrates placed on the bottom).

Some modifications resulted in a global biological index (GBI, 1985) and a standardized global biological index (SGBI, AFNOR, 1992). The latter has tended to replace the earlier indexes (Table 29).

The SGBI is based on a list of 138 taxa, of which 38 are bioindicators. Eight ranges of samples are taken at each point as a function of the current, substrate, and vegetation present. The table of biotic indexes (Table 29) comprises 9 groups of bioindicators, 14 classes of taxonomic richness, and 20 classes of indexes. The SGBI cannot be used at sources (crenal), certain lower courses of rivers, and in atypical environments such as canals and estuary zones (hypopotamon).

3. Conclusions

Analysis of the physicochemical parameters of a stretch of water gives only an instantaneous indication of its quality. It does not take into account variations of this state over time, especially after intermittent polluting effluents. For that, a permanent and costly record of parameters would be necessary.

Biological methods of evaluating pollution with recourse to biochemistry and ecotoxicology measure either the trophic state of an environment or the degree of species sensitivity to pollution.

Table 27. Tuffery and Verneaux's biotic indexes (1967)

Faunal groups		Total number of taxonomic units (TU) present				
		0–1	2–5	6–10	11–15	> 15
		Biotic index				
1. Plecoptera or	> 1 TU	–	7	8	9	10
Ecdyonuridae	1 TU	5	6	7	8	9
2. Trichoptera	> 1 TU	–	6	7	8	9
with sheath	1 TU	5	5	6	7	8
3. Ancylidae	> 2 TU	–	5	6	7	8
Ephemeroptera except Ecdyonuridae	1 or 2 TU	3	4	5	6	7
4. Aphelocheirus Odonata, Gammaridae or Mollusca (except Sphaeridae)	All TU absent	3	4	5	6	7
5. Asellus, Hirudinea, Sphaeridae or Hemipterae (except Aphelocheirus)	All TU absent	2	3	4	5	–
6. Tubificidae or Chironomidae *thumni-plumosus*	All TU absent	1	2	3	–	–
7. Eristalinae	All TU absent	0	1	1	–	/

Table 28. Practical limits of taxonomic identification for calculation of Tuffery and Verneaux's biotic index (1967)

Orders	Limits	Orders	Limits
Plecoptera	genus	Hemiptera	genus
Trichoptera	family	Diptera	family, sub-family, or tribe*
Ephemeroptera	genus	Turbellaria	genus or species*
Odonata	genus	Hirudinea	genus or species*
Coleoptera	family	Oligochaeta	family
Mollusca	genus or species*	Nematodes	presence
Crustacea	family	Hydracarids	presence
Megaloptera	genus		

*Depending on the case.

Table 29. Identification table for standardized global biological index (AFNOR, 1992)

	Class of variety	14	13	12	11	10	9	8	7	6	5	4	3	2	1
	Number of taxa	> 50	49 45	44 41	40 37	36 33	32 29	28 25	24 21	20 17	16 13	12 10	9 7	6 4	3 1
9	Chloroperlidae	20	20	20	19	18	17	16	15	14	13	12	11	10	9
	Perlidae														
	Perlodidae														
	Taeniopterygiidae														
8	Capniidae	20	20	19	18	17	16	15	14	13	12	11	10	9	8
	Brachycentridae														
	Ontoceridae														
	Philopotamidae														
7	Leuctridae	20	19	18	17	16	15	14	13	12	11	10	9	8	7
	Glossosomatidae														
	Beraeidae														
	Goeridae														
	Leptophlebiidae														
6	Nemouridae	19	18	17	16	15	14	13	12	11	10	9	8	7	6
	Lepidostomatidae														
	Sericostromatidae														
	Ephemeridae														
5	Hydroptilidae	18	17	16	15	14	13	12	11	10	9	8	7	6	5
	Heptageniidae														
	Polymitarcidae														
	Pothamantidae														
4	Leptoceridae	17	16	15	14	13	12	11	10	9	8	7	6	5	4
	Polycentropidae														
	Psychomiidae														
	Rhyacophilidae														
3	Limnephilidae*	16	15	14	13	12	11	10	9	8	7	6	5	4	3
	Hydropsychidae														
	Ephemerellidae*														
	Aphelocheiridae														
2	Baetidae*	15	14	13	12	11	10	9	8	7	6	5	4	3	2
	Caenidae*														
	Elmidae*														
	Gammaridae*														
	Mollusca														
1	Chironomidae*	14	13	12	11	10	9	8	7	6	5	4	3	2	1
	Asellidae*														
	Achaeta														
	Oligochaeta*														

*Taxa represented by at least 10 individuals. The remainder are represented by at least 3 individuals.

Unlike in chemical analysis, benthic communities assimilate the qualities of the water and sediments over long periods ranging from weeks (algae) to months (macrophytes and macro-invertebrates). The organisms are records of temporary pollution. They may also behave as cumulative receivers, by concentrating pollutants in their tissues at concentrations in water lower than the sensitivity threshold of chemical methods of analysis. Benthic communities are permanent observatories of the environment, and their study gives a fairly accurate idea of the quality of the water and substrate. However, it is not the nature of the pollution that is detected, but only the intensity of its manifestations in the life of organisms and structure of communities. Biotic indexes are the synthetic expression of the general quality of an environment, all causes considered.

The fact that there are about 50 methods of comparative analysis of communities (diversity index, similarity index, multivariable analyses, etc.) and, for Europe alone, about 40 methods that use biological indicators, shows that there is no ideal or universal method. There are several reasons for this.

The typology of pollution from biological indicators and biotic indexes is not independent of the natural typology of running water, based on hydraulics (Illies and Botosaneanu's zonation in Europe). One or the other is dominant, depending on environmental conditions. An identical organic load has different effects on a rhithral, with a rapid current and oxygenation facilitated by turbulence and shallow depth, than on a potamal. In the first case, factors appropriate to natural zonation are most often dominant. In the second case, the organic load has a primordial effect. In the rhithral, a slight organic pollution favours the development of pollution-resistant detritivorous species without eliminating pollution-sensitive species. Biotic indexes are a poor reflection of the actual quality of the environment.

Communities of the potamal are different from those of the rhithral. In a rhithral, an SGBI of 17 is already a sign of a slight alteration in the environment. The same index in the potamal reflects an excellent water and substrate quality, because the indicator species of groups 8 and 9 (Table 29), which are characteristic of the rhithral, have disappeared. Consequently, an ambiguity exists in biotic indexes, the communities reflecting the natural typological level as well as alterations in the environment, particularly at the transition point between rhithral and potamal.

The geographical distribution of organisms limits the use of indicator species in a given geographical area. Species found in the Scandinavian countries or England, for example, are not the same as those found in Southern Europe. The same taxa can have different indicator values depending on the regions. The genera *Baetis* (Ephemeroptera) and *Nemoura* (Plecoptera), for example, which are highly sensitive to pollution in Great Britain, are more tolerant in Denmark. Biotic indexes derived from the index measured on the Trent have consequently been adapted for use in Denmark, Germany, Belgium, France, Spain, and Portugal. The principle of the method is universal, but the indicator taxa are not. Their value is limited to a given ecological region, an *ecoregion*. This explains the large number of biotic indexes on a single continent such as Europe. The methods taking into account the structure of communities are nearly identical in homologous ecosystems, but with different taxa presenting the same type of reaction with respect to the same type of pollution.

The important thing is not the multiplicity of methods (since no single method is universal), but the extent of their agreement, even if their sensitivity is different. This consistency has been verified for various methods (Fig. 63).

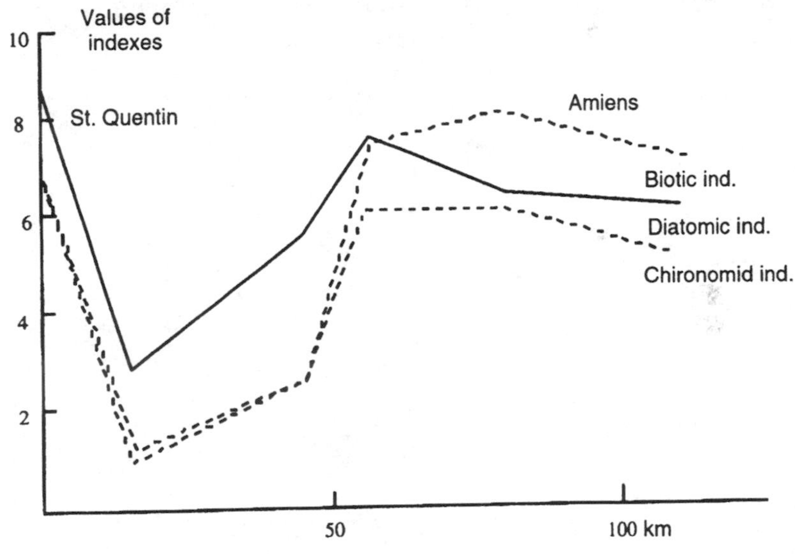

Fig. 63. Average values of biotic indexes (Tuffery and Verneaux), diatomic index, and chironomid index on the Somme in France, from the source to downstream of Amiens (Bazerque, Laville and Brouquet, 1989)

To all these problems is added that of sampling. Indexes based on Diatoms, for example, require regular samples, because their biological cycle is short. The macro-invertebrate cycle extends over several months, and two ranges of samples a year suffice to estimate the quality of an environment.

On the other hand, the aim is to obtain a global pollution index from samples taken from a mosaic of microhabitats. Tuffery and Verneaux's index is calculated from just two samples (in slow and rapid facies, and in lentic and lotic facies), the GIBQ from six current-substrate pairs, and the SGBI from eight, for an equivalent area (20×10^{-3} m^2). The number of habitats varies from one site to another; it is higher in a rhithral than in a potamal. Corrections must be made for uniform beds (e.g., straightened courses, mud beds). The methods of evaluating pollution become more delicate as they become more sensitive. For each, its reliability and limitations must be known, as well as its agreement with other methods.

Quick methods are sufficient to evaluate major pollution such as that of the Riou-Mort, i.e., percentage of Tubificidae in relation to the total number of Oligochaeta, or even of macro-invertebrates. Tuffery and Verneaux's index gives sufficient evaluations in most cases of pollution, and it is only the most subtle types of pollution that call for an index such as the SGBI. The most sophisticated biotic indexes perhaps contribute more at the fundamental level than at the applied level.

No matter what means are used, once the extent of pollution has been estimated, what is then required is knowledge of its cause (by methods of chemical analysis) so that it can be eliminated.

17

Conclusions

To understand the ecology of running waters, a concept from physics has to be adopted: the concept of field, a demarcated space in which there is an organizing factor. In running water, the field is the bed of the water course, from the source to the confluence or estuary on a longitudinal profile. The organizing factor is the current, which has a three-fold action:

— as an agent of erosion, transport, and redistribution of materials from the watershed;
— as an agent of erosion, transport, and redistribution of benthic organisms (drift);
— as an agent of transport of dissolved substances, including domestic and industrial waste, and organisms in suspension (plankton).

Everything that lives in the water course, from bacteria to fish, is subjected to the current, either directly by its erosive action or indirectly by the intermediary role of transit time.

The erosion, transport, and redistribution of materials from the earth's crust involves minerals (pebbles, gravel) and organic matter from the watershed. In each part of the bed, the force and competence of the current determine the nature of the substrate, its granulometry. On steep slopes, water flows in many directions, and the result is a mosaic of microhabitats. When the slope diminishes, the flow is mainly unidirectional. The types of microhabitats on the profile are different, and they cover a larger area. The way an ecosystem is perceived depends on the scale of observation: a mosaic of microhabitats on a small scale or an upstream-downstream continuum (with or without nodas) on a large scale.

The hydrological regime, with an annual succession of high and low water periods and a current with variable force and competence, along with the hydraulic factor, determines a chronic instability of the superficial horizon of the substrate. The annual hydrological cycle also determines interactions between the alluvial flood plain and the water course.

Organisms that colonize the superficial horizon of the substrate (benthos) have a strategy of adjustment to the morphodynamic conditions of the water. Using this strategy they aim to maintain themselves in their habitat despite the current, which tends to want to carry them downstream. This strategy manifests itself in various forms by:

— adherence to the substrate by means of mucus, hooks, or suckers;
— high reproductive rates, compensating for losses due to drift;

— a flying phase in the biological cycle, allowing constant recolonization of substrates (as insects with an aquatic larval phase and an adult flying phase). Torrential zones are characterized by the dominance of insects that have such a strategy (especially Ephemeroptera, Plecoptera, Trichoptera, and Diptera).

Two environments, however, escape this chronic instability: the hyporheic and that constituted by macrophytes, particularly Bryophytes. The latter are fixed onto the mother rock or stable stones. They offer a perennial environment for algae and fauna, one sheltered from the current, in the sense that the species involved are small or at juvenile stages. Invertebrates without a flying adult phase—especially Crustacea and Oligochaeta—are thus abundant.

The transport and redistribution of organic matter from the watershed poses another problem, that of energy flow and flow of nutrients in the ecosystem. In terrestrial, oceanic, and lacustrine ecosystems, the biosynthesis-biodegradation gradient is vertical. All dead matter—the necromass—falls on the soil or the lake or sea bed and is degraded and mineralized by fungi and bacteria. In running waters, the gradient is longitudinal, upstream to downstream. The original source of the energy and nutrients needed for the ecosystem to function is the necromass of the watershed, carried by runoff. The running water ecosystem does not begin with the transformation of light energy into chemical energy by photosynthesis, but with the utilization of the necromass. It is related to a terrestrial soil or ocean or lake bed and is heterotrophic (P/R ratio < 1). The mineral salts resulting from biodegradation are at the origin of algal and macrophyte synthesis downstream, particularly nitrogen and phosphorus salts, which are rare or absent in the rocks of the watershed. From upstream to downstream, the ecosystem tends towards autotrophy ($P/R > 1$).

In low valleys, the alluvial flood plain constitutes a reserve of organic matter (such as leaves) that is suspended and transported by floods. Its decomposition then leads to the reversion of the water course to heterotrophy.

Plankton in suspension appears at low water in the potamal and flows with the water. The limiting factor of its development is thus water transit time, the time it takes to flow down to the confluence or estuary. This transit time, brief in comparison to that of a lake, permits only the development of algae or invertebrates with a rapid rate of multiplication, similar to Rotifers. Relatively long transit times are required for the predator-prey relationship or browsing by Rotifers to play a significant role in regulating populations of planktonic algae. Such long transit times are seen, for example, in the Seine from Montereau downstream.

The evolution of the running water ecosystem from the pioneer stage to a climax stage does not occur over time. It is a longitudinal evolution, from upstream to downstream, linked to the evolution of the hydraulic factor (slope, current, flow) and temperature. This hydraulic factor determines the longitudinal zonation of benthos and the possibilities of plankton development. Pioneer organisms and ecosystems are characterized by the predominance of abiotic factors, which confirms the major role of seeding: recolonization of the upstream sector from eggs laid by insects with flying adults and plankton seeding. Under these conditions, the explicative value of knowing the energy and matter flows in the ecosystem may be secondary, just as with a pioneer stage on bare soil, where seeding and climatic conditions play a dominant role. The evaluations by various authors of energy and matter flows in running waters constitute an interesting input in terms of knowledge, but contribute little to the understanding of the functioning of an ecosystem regulated for the most part by abiotic factors such as hydraulics and temperature. These are the two factors that are modified by human activity in developing water courses.

Modifications carried out on rivers until the 19th century were essentially designed to promote navigation, recover farm land from the flood plain, and protect settlements in the lower valleys against flood. To facilitate navigation, the water level was raised by navigation embankments (in this way a depth of 1.10 to 2.50 m was obtained on the Lot, for example). The potamal was thus extended upstream, at the expense of the rhithral. The enlargement of the submerged section of the bed, with an unchanged flow, had the effect of increasing transit time and raising the summer temperature. This type of expansion in fact prepared the conditions that led to eutrophication of rivers in the 20th century. The recovery of farm land by the construction of dykes eliminated or reduced the flood plain, which was a source of nutrients leached into the soil during floods.

Developments in the 20th century involved mostly hydroelectric and irrigation dams. In the low valleys, low-volume dams were introduced in the river simply to create a higher fall. Like the navigation embankments, they increase water transit time. In the middle and upper valleys, the rhithral, and in enclosed valleys, the dams have a large storage capacity. They retain water from the watershed during high water and gradually release it during low water. The water regime is thus modified by the control of floods and the maintenance of flow during the natural low water period. Its temperature is also affected by storage of cold winter or spring waters.

The power of a hydroelectric plant being proportionate to the flow and fall height, these two factors are increased by diverting the water, at the foot of the reservoirs, by means of pressure pipelines that restore the water to the river downstream, at a lower altitude. A sector of the river is thus bypassed, and only a slow flow remains in the bed (called the reserve flow). The biota is modified, the low current transforming an epirhithron, for example, into a meta- or hyporhithron. The intermittent functioning of hydroelectric plants (at peak hours of consumption) has led to a sluice-dominated regime—sudden increase in flow at certain hours of the day. In a rhithral, the drift of organisms increases at the beginning of a sluice. In a potamal, the fine organic and clay sediments return to a suspended state with each sluicing. This is a phenomenon analogous to that observed in canals (every time a barge passes, the sediments are re-suspended) or in a tidal rocking zone in estuaries. This phenomenon is found particularly in rivers with navigation embankments, in which the finest sediments are deposited (for example, the Lot or the Vézère in France). Once the sediments are re-suspended, there is not enough time for sedimentation between two sluice gates, and the result is permanent turbidity.

Since the second half of the 19th century, the growth of towns and cities and industrial development have led to the use of rivers and streams to drain away waste products. Mineral nitrous and phosphate wastes contribute to plant nutrition (of algae and macrophytes) and the eutrophication of waters. The same is true of organic wastes, after degradation and mineralization. The conditions of eutrophication, however, remain largely controlled by current speed via the intermediary of transit time in the potamal and erosion of phytobenthos in the rhithral. For short transit times (flows in the Lot higher than 20 m^3/s, for example), current speed continues to regulate phytoplankton growth. When transit time increases, mineral nitrogen and phosphorus regulate algal development. For long transit times (lower than 20 m^3/s in the Lot) and very high nitrogen and especially phosphorus concentrations, light and temperature control the algal production and thus become the amplifying factors.

The degeneration phases of algae (Cyanophyceae in particular) contribute necromass to the water course that will be mineralized and re-used by the phytoplankton and phytobenthos. Organic wastes (domestic and industrial) are also oxidized and mineralized. Mineral and organic inputs constitute two aspects of a single problem—that of enrichment of the energy potential of the ecosystem, or eutrophication. It represents an immediate enrichment by mineral input, which differs from organic nutrient input. The transit time determines the possibilities of the use of this potential by the phytoplankton.

In the potamal, the excess of organic matter can lead to a fall in dissolved oxygen concentrations. Real pollution, no longer simply trophic, thus appears, by asphyxia of organisms and toxic wastes from fermentation (e.g., ammonia, hydrogen sulphide). As for non-biodegradable wastes (chemical wastes, heavy metals), their toxicity is variable, depending on their chemical constitution, their concentration, and the possibility of their accumulation in the tissues of organisms. Each taxon has a specific reaction to each toxin and its own limits of tolerance.

According to B. Statzner et al. (1986, 1988), hydraulics and hydrology are the key factors in the ecology of running waters. The temperature, estimated in annual degree-days, depends partly on the water regime and hydrology, apart from the seasonal rhythm. In Europe, Illies and Botosaneanu's zonation, in 1963, was based on hydraulics and temperature. The variability of these two factors and the pertubations they cause in the environment belie the dogma of the stability and equilibrium of ecosystems that has long prevailed in ecology. The predominance of physical factors maintains lotic ecosystems in the pioneer stage. This is the major role of hydraulics—current speed and flow—and it is this that makes these moving paths, the running waters, a singular chapter in the ecology of continental waters.

In stagnant waters, the influence of hydraulic factors and hydrology disappears, whereas transit time, which is very long, becomes a secondary ecological factor. Along with the temperature factor, light, dissolved salts, energy flow, and cycling time of matter become predominant, especially with respect to problems of eutrophication.

Bibliography

Anon. 1988. Les poissons et leur environnement dans les grands fleuves européens. *Sciences de l'Eau*, 7, 1: 1–154.

Agences de l'Eau. 1993. *Étude bibliographique des méthodes biologiques d'évaluation de la qualité des eaux de surface continentales*. Étude no. 35, 3 vol.

Allan, J.D. 1994. *Stream Ecology*. Chapman & Hall, London.

Amiard-Triquet, C.A., and Amiard, J.C. 1980. *Radioécologie des Milieux Aquatiques*. Masson, Paris.

Amoros, C., and Petts, G.E., ed. 1993. *Hydrosystèmes Fluviaux*. Masson, Paris.

Anderson, M.G., and Burt, T.P. 1985. *Hydrological Forecasting*. John Wiley & Sons, Chichester.

Armantrout, N.B., ed. 1981. *Acquisition and Utilization of Aquatic Habitat. Inventory Information*. Amer. Fisheries Soc., Western Division, Bethesda, Maryland.

Arrignon, J. 1998. *Aménagement Piscicole des Eaux Douces*. Éditions Tec & Doc, Paris.

Bagliniere, J.L., and Maisse, G., ed. 1991. *La Truite. Biologie et Écologie*. INRA, Paris.

Balvay, G., ed. 1995. *Space Partition within Aquatic Ecosystems*. Proc. 2nd Intern. Congr. of Limnology. Kluwer Acad. Publishers, Dordrecht.

Banarescu, P. 1990–1995. *Zoogeography of Freshwater*, 3 vol. Aula-Verlag, Wiesbaden.

Barnes, J.R., and Minshall, G.W. 1983. *Stream Ecology. Application and Testing of General Ecological Theory*. Plenum Press, New York.

Bartram, J., and Ballance, R. 1996. *Water Quality Monitoring: A Practical Guide to the Design and Implementation of Freshwater Quality Studying and Monitoring Programmes*. Chapman & Hall, London.

Baumgartner, A., and Liebscher, H.J. 1996. *Allgemeine Hydrologie. Quantitative Hydrology*. Gebrüder Borntraeger, Berlin.

Baumgartner, A., and Reichel, E. 1975. *The World Water Balance*. Elsevier Sci., Amsterdam.

Behning, A. 1928. *Das Leben der Volga*. Die Binnengewässer, V. Schweizerbart'sche Verlag, Stuttgart.

Bertrand, H. 1954. *Les Insectes Aquatiques d'Europe*, 2 vol. P. Lechevalier, Paris.

Bevan, K., and Carling, P., ed. 1989. *Floods: Hydrological, Sedimentological and Geomorphological Implications*. John Wiley & Sons, Chichester.

Board, R.G., and Lovelock, D.W. 1973. *Sampling Microbial Monitoring of Environments*. Academic Press, London.

Bone, Q., Marshall, N.E., and Blaxter, J.H.S. 1994. *Biology of Fishes*. Chapman & Hall, London.

Bonin, D.J., and Golterman, J.H.S., 1994. *Flux between Trophic Levels and through the Water Sediment*. Development in Hydrobiology. Kluwer Acad. Publ., Dordrecht.

Boon, P.J., Calow, P., and Petts, G.E., ed. 1992. *River Conservation and Management*. John Wiley & Sons, Chichester.

Boon, P.J., and Howell, D.L., ed. 1997. *Freshwater Quality: Defining the Indefinable*? Scottish Publ. Sak., Edinburgh.

Botosaneanu, L., ed. 1998. *Studies in Crenobiology. The Biology of Springs and Springbrooks*. Backhuys Publ., Leiden.

Braukmann, U. 1987. *Zoozönologische und saprobiologische Beiträge zu einer allgemeinen regionale Bachtypologie*. Ergebnisse der Limnologie, 26. E. Schweizerbart'sche Verlag, Stuttgart.

Bravard, J.P. 1987. *Le Rhône, du Léman à Lyon*. La Manufacture, Lyon.

Bretschko, G., and Helesic, J., ed. 1998. *Advances in River Bottom Ecology*. Backhuys Publ., Leiden.

Brookes, K. 1988. *Channelized Rivers: Perspectives for Environmental Management*. John Wiley & Sons, Chichester.

Brookes, K., and Shields, F.D. 1998. *River Channel Restoration*. J. Wiley & Sons, Chichester.

Cairns, J. Jr. 1977. *Aquatic Microbial Communities*. Garland Publ. Inc., New York and London.

Cairns, J. Jr. 1992. *Restoration of Aquatic Ecosystems*. John Wiley & Sons, Chichester.

Calow, P., and Petts, G.E., ed. 1994. *Rivers Handbook*, 2 vol. Blackwell Scient. Publ., Oxford.

Campbell, L.I.C., ed. 1990. *Mayflies and Stoneflies: Life History and Biology*. Kluwer Acad. Publ., Dordrecht.

Carling, P.A., and Petts, G.E., ed. 1992. *Lowland Floodplain Rivers: Geomorphological Perspectives*. John Wiley & Sons, Chichester.

CEBEDEAU. 1996. *Le Livre de l'Eau*. CEBEDAU, Liège.

CEMAGREF. 1991. *L'Eutrophisation du Fleuve Charente*. Agence de l'Eau Adour-Garonne, Toulouse. Mimeo.

Champart, D., and Larpent, J.P., ed. 1988. *Biologie des Eaux*. Masson, Paris.

Chapman, D., ed. 1996. *Water Quality Assessment. A Guide to the Use of Biota, Sediments and Water in Environmental Monitoring*. Chapman & Hall, London.

Chow, V.T., ed. 1964. *Handbook of Applied Hydrology*. McGraw-Hill, New York.

Ciaccio, L.L., ed. 1971. *Water and Water Pollution Handbook*. M. Dekker, New York.

Collinson, J.D., and Lewin, J., ed. 1983. *Modern and Ancient Fluvial Systems*. Blackwell Scient. Publ., Oxford.

Cook, C.D.K. 1996. *Aquatic Plant Book*. SPB Academic Publ., Amsterdam.

Cosgrove, D.E.. and Petts, G.E., ed. 1990. *Water Engineering and Landscape*. Belhaven, London.

Craig, J.F., and Kemper, J.B., ed. 1987. *Regulated Streams*. Plenum Press, New York.

Crickway, C.H. 1975. *The Work of the Rivers. A Critical Study of the Central Aspects of Geomorphology*. Macmillan, London.

Cummins, K.W. 1996. *Lotic Limnology*. Chapman & Hall, London.

Cushing, C.E., Cummins, K.W., and Minshall, G.W., ed. 1995. *River and Stream Ecosystems*. Ecosystems of the World, 22. Elsevier Science, Amsterdam.

Davis, B.R., and Walker, K.F., ed. 1986. *The Ecology of River Systems*. W. Junk Publ., The Hague.

De Dekker, P., and Williams, W.D. 1986. *Limnology in Australia*. W. Junk Publ., The Hague.

Degens, E.T., Kempe, S., and Richey, J.E., ed. 1991. *Biochemistry of Major World Rivers*. John Wiley & Sons, Chichester.

Descy, J.P., Reynolds, C.S., and Padisak, J., ed. 1994. *Phytoplankton in Turbid Environments: Rivers and Shallow Lakes*. Kluwer Acad. Publ., Dordrecht.

Dodge, J.P., ed. 1989. *Proceeding of the International Large River Symposium*. Canadian Spec. Publ. of Fisheries and Aquatic Sciences, 106, Montreal.

Dudgeon, D. 1992 *Patterns and Process in Stream Ecology. A Synoptic Review of Hong Kong Running Waters*. Die Binnengewässer, 29. E. Schweizerbart' sche Verl. Stuttgart.

Dudley, W. 1987. *The Ecology of Temporary Waters*. Croom Helm, London.

Dussart, B. 1992. *Limnologie. L'étude des Eaux Continentales*. 2nd ed., Boubée, Paris.

Edeline, F. 1994 *L'Èpuration Biologique des Eaux*. Cébedoc, Brusells.

Edwards, R.W., and Brooker, M.P. 1982. *The Ecology of the River Wye*. Monographiae Biologicae, 50. W. Junk, The Hague.

Elliot, C.M. 1984. *River Meandering*. Amer. Soc. of Civil Engineers, New York.

Elliott, J.M. 1994. *Quantitative Ecology and Brown Trout*. Oxford University Press, Corby.

ENPC. 1998. *Les Systèmes Fluviaux Anthropisés*. Presses de l'École Nationale des Ponts et Chaussées, Paris.

Ethridge, F.G., Flores, R.H., and Harvey M.D., ed. 1987. *Recent Developments in Fluvial Sedimentology*. Soc. Economic Paleontologists and Mineralogists, Publ. no. 39, Oklahoma.

Fontaine, T.D., and Bartell, S.M., ed. 1983. *Dynamic of Lotic Ecosystems*. Ann Arbor Sci., Ann Arbor, Michigan.

Ford, T.E., ed. 1993. *Aquatic Microbiology. An Ecological Approach*. Blackwell Scient. Publ., London.

Förstner, U., and Wittmann, G.T.W., ed. 1979. *Metal Pollution in the Aquatic Environment*. Springer-Verlag, Berlin.

Foster, I.D.L., Gurnell, A.M., and Webb, B., ed. 1995. *Sediment and Water Quality in River Channel*. John Wiley and Sons, Chichester.

Frecaut, R., and Pagney, P. 1983. *Dynamique des Climats et de l'Écoulement Fluvial*. Masson, Paris.

Fustec, E., and De Marsiley, G., ed. 1993. *La Seine et son Bassin: de la Recherche à la Gestion*. Actes du Colloque PIREN-Seine, 29–30 April 1993. Paris.

Gaujous, D. 1995. *La Pollution des Milieux Aquatiques*. Tec et Doc Laviosier, Paris.

Gehm, H.W., and Bregmans, J.I., ed. 1976. *Water Resources and Pollution Control*. Van Nostrand Reinhold, New York.

Gessner, F. 1955–1959. *Hydrobotanik*, Vol. 1, 1955. Vol. 2, 1959. Veb. Deutscher Verlag der Wissenschaften, Berlin.

Gibert, J., Danielopol, D.L., Stanford, J., and Thorp, J.H., ed. 1994. *Groundwater Ecology*. Academic Press, London.

Ginet R., and Decou, V. 1977. *Initiation à la Biologie et à l'Écologie Souterraines*. Delarge, Paris.

Goltermann, H.L., and Clino, R.S., ed. 1967. *Chemical Environment in the Aquatic Habitat*. IBP Symposium, Amsterdam et Nieuerluis. N.W. Nord-Holland, Amsterdam.

Goltermann, H.L., Sly, P.G., and Thomas, R.L. 1983. *Study of the Relationship between Water Quality and Sediment Transport*. Technical papers in Hydrology, no. 26. UNESCO, Paris.

Gopal, B., and Wetzel, R.G. ed. 1995. *Limnology in Developing Countries*. International Scientific Publications, New Delhi.

Gordon, N.D., McMohan, T.A., and Finlayson, B.L. 1992. *Stream Hydrology. An Introduction for Ecologists*. John Wiley & Sons, Chichester.

Gore, J.A., ed. 1985. *The Restoration of Rivers and Streams*. Butterwort Publ., Boston.

Gore, J.A., and Petts G.E., ed. 1989. *Alternatives in Regulated Rivers Management*. CRC Press, Boca Raton, Florida.

Gregory, K.J., and Walling, D.E. 1974. *Drainage Basin: Form and Process*. Edward Arnold, London.

Guilcher, A. 1979. *Précis d'Hydrologie Marine et Continentale*. Masson, Paris.

Gurnell A.M., and Petts, G.E., ed. 1995. *Changing River Channels*. John Wiley & Sons, Chichester.

Hakanson, L., 1999. *Water Pollution*. Backhuys Publ., Leiden.

Hakanson, L., and Peters, P.H. 1995. *Predictive Limnology. Methods for Predictive Modelling*. SPB Academic Publ., Amsterdam.

Hart, C.W., and Fuller, S.L.H., ed. 1974. *Pollution Ecology of Freshwater Invertebrates*. Academic Press, New York.

Haslam, S.M. 1978. *River Plants: the Macrophytic Vegetation of Watercourse*. Cambridge Univ. Press, Cambridge.

Haslam, S.M. 1992. *River Pollution. An Ecological Perspective*. Cambridge Univ. Press, Cambridge.

Haslam S.M., and Wolseley, P.A. 1981. *River Vegetation. Its Identification, Assessment and Management*. Cambridge Univ. Press, Cambridge.

Haslam S.M., and Wolseley, P.A. 1987. *River Plants of Western Europe. The Macrophytic Vegetation of the European Economic Community*. Cambridge Univ. Press., Cambridge.

Hellawell, J.M. 1978. *Biological Surveillance of Rivers: a Biological Monitoring Handbook*. Water Res. Center, Medmenhans.

Hellawell, J.M. 1986. *Biological Indicator of Freshwater Environmental Management*. Elsevier, Amsterdam.

Hemphill, R.W., and Bramley, M.E. 1989. *Protection of River and Canal Banks*. CIRIA, Butterworths, London.

Herschy, R.W., and Fairbridge, R.W. ed. 1998. *Encyclopedia of Hydrology and Water Resources*. Kluwer Acad. Publ., Dordrecht.

Hey, R.D., Bathurst, J.C., and Thorne, C.R., ed. 1982. *Gravel-Bed Rivers*. John Wiley & Sons, Chichester.

Holcik, J., ed. 1989. *The Freshwater Fishes of Europe*. Aula-Verlag, Wiesbaden.

Huet, M. 1970. *Traité de Pisciculture*. De Wyngaert, Brussels.

Hutchinson, G.E. 1975–1993. *A Treatise on Limnology. I: Geography and Physics of Lakes*, 1975. II: *An Introduction to Lake Biology and Limnoplankton*, 1996. III: *Limnological Botany*, 1975. IV: *The Zoobenthos*, 1993. John Wiley & Sons, Chichester.

Hynes, H.B.N. 1960. *The Biology of Polluted Waters*. Liverpool Univ. Press, Liverpool.

Hynes, H.B.N. 1970. *The Ecology of Running Waters*. Liverpool Univ. Press, Liverpool.

Illies J., ed. 1978. *Limnofauna Europaea*. G. Fisher, Iéna.

Illies, J., and Botosaneanu, L., 1963. *Problèmes et Méthodes de la Classification et de la Zonation Écologique des Eaux Courantes Considérees surtout du Point de Vue Faunistique*. Schweizerbart'sche Verlag, Stuttgart.

Inhaber, H. 1976. *Environmental Indices*. John Wiley & Sons, Chichester.

James, A., and Eviron, L.E., ed. 1979. *Biological Indicators of Water Quality*. John Wiley & Sons, Chichester.

Jeffrey, D.W., and Madden, B., ed. 1991. *Bioindicators and Environmental Management*. Academic Press, London.

Jenkins, S.J., ed. 1973. *Advances in Water Pollution Research*. Pergamon Press, London.

Johnson, D.W., and Van Hook, R.I. 1989. *Analysis of Biogeochemical Cycling Processes in Walker Branch Watershed*. Springer-Verlag, Berlin.

Karr, J.R., and Schlosser, I.J. 1996. *Landscape Disturbance and Stream Ecosystems*. Chapman & Hall, London.

Kaush, H., and Michaelis, W., ed. 1996. *Suspended Particulate Matter in Rivers and Estuaries*. Schweizerbart'sche Verlag, Stuttgart.

Kempe, S., and Richey, J.E., ed. 1991. *Biogeochemistry of Major World Rivers*. John Wiley & Sons, Chichester.

Kemper, J.B., and Graig, J., ed. 1987. *Regulated Streams. Advances in Ecology*. Plenum Press, New York.

Kerfood, W.C., and Sih, A., ed. 1987. *Predation. Direct and Indirect Impacts on Aquatic Communities*. University Press of New England, Hanover and London.

Kinzelbach, H.R., and Friedrich, G. 1990. *Biologie des Rheins*. Limnologie aktuell, l. Schwezerbart'sche Verlag, Stuttgart.

Klapper, H. 1992. *Control of Eutrophisation in Inland Waters*. Ellis Horwood Ltd., London.

Klein, L. 1962. *River Pollution*, 3 vol. Butterworths, London.

Knighton, A.D. 1984. *Fluvial Forms and Processes*. Edward Arnold, London.

Lacaze, J.C. 1996. *L'Eutrophisation des Eaux Marines et Continentales*. Ellipses, Paris.

Lambert, R. 1996. *Géographie du Cycle de l'Eau*. Presses de l'Université du Mirail, Toulouse.

Lamotte, M., and Bourlière, F., ed. 1971. *Problèmes d'Écologie: l'Échantillonage des Peuplements Animaux des Milieux Aquatiques*. Masson, Paris.

Lamotte, M., and Bourlière F., ed. 1983. *Structure et Fonctionnement des Écosystèmes Limniques*. Masson, Paris.

Lampert, W., and Sommer, U. 1997. *Limnoecology*. Oxford University Press, Gorby.

Langford, T.E.L. 1996. *Biology of Pollution in Major River Systems*. Chapman & Hall, London.

Larras, J. 1977. *Fleuves et Rivières non Aménagés*. Eyrolles, Paris.

Le Cren, E.D., and McConnel, R.H., ed. 1980. *The Functioning of Freshwater Ecosystems*. PBI, Cambridge Univ. Press, Cambridge.

Leopold, L.B. Wolman, M.G., and Mille, J.P. 1964. *Fluvial Process in Geomorphology*. Freeman, San Francisco.

Lewin, J., ed. 1981. *British Rivers*. George Allen & Unwin, London.

Liebman, H. 1962. *Handbuch der Frischwasser und Abwasserbiologie*. G. Fischer, Iéna.

Liepolt, R., Ed. 1967. *Limnologie der Donau*. Schweizerbart'sche Verlag, Stuttgart.

Lillehammer, A., and Saltveit, S.J. 1984. *Regulated Rivers*. Universitetsforlaget, Oslo.

Lockwood, A.M.P. 1976. *Effects of Pollutants on Aquatic Organisms*. Cambridge University Press, Cambridge.

Loock, M.A., and Williams, D.D., ed. 1981. *Perspectives in Running Water Ecology*. Plenum Press. New York.

Luczkwitch, J.J., Motta, P.J., Norton, S.F., and Liem, K.F. 1995. *Ecomorphology of Fishes*. Kluwer Acad. Publ., Dordrecht.

McCafferty, W.P. 1983. *Aquatic Entomology*. 2nd ed. Jones & Barlett Publ. Inc., USA.

Macan, T.T., and Worthington, E.B. 1968. *Life in Lakes and Rivers*. 2nd ed. Collins Clear Type Press, London.

Maitland, P.S., and Morgan, N.C. 1997. *Conservation Management of Freshwater Habitats*. Chapman & Hall, London.

Mance, G. 1987. *Pollution Threat of Heavy Metals in Aquatic Environment*. Elsevier, Amsterdam.

Margalef, R. 1983. *Limnologia*. Edic. Omega, Barcelone.

Margalef, R., ed. 1994. *Limnology Now. A Paradigm of Planetary Problems*. Elsevier, Amsterdam.

Matthew, W.J. 1996. *Patterns in Freshwater Fish Ecology*. Chapman & Hall, London.

Matthews, W.J., and Heims, D.C. 1987. *Community and Evolutionary Ecology of North American Stream Fishes*. University of Oklahoma Press, Normon.

Merritt, R.W., and Cummins, K.W., 1978. *An Introduction to the Aquatic Insects of North America*. Kendall-Hunt Publ. Company, USA.

Meybeck, M., Chapman, D., and Helmer, R. 1989. *Global Freshwater Quality: a First Assessment. Global Environment Monitoring System*. Blackwell Sci. Publ., Oxford.

Mills, D., ed. 1992. *Strategies for the Rehabilitation of Salmon Rivers*. Inst. of Fisheries Management, London.

Minshall, G.W., and Barnes R., ed. 1983. *Stream Ecology. Application and Testing of General Theory*. Plenum Press, New York.

Mitchell, R., ed. 1972. *Water Pollution Microbiology*. Wiley Interscience, New York.

Mordukkai-Boltovskoi, P.D., ed. 1979. *The River Volga and Its Life*. W. Junk, The Hague.

Morgan, N.C., and Maitland, P.S. 1997. *Conservation Management of Freshwater Habitats*. Kluwer Academic Publ., Dordrecht.

Morisawa, M. 1968. *Streams: Their Dynamics and Morphology*. McGraw-Hill Co., London.

Morisawa, M. 1973. *Fluvial Sedimentology*. G. Allen & Unwin, London.

Morisawa, M. 1985. *Rivers, Form and Process*. Longman, London.

Moss, B. 1984. *Ecology of Freshwaters: Man and Medium*. Blackwell Scient. Publ., Oxford.

Naiman, R.J., and Decamps. H., ed. 1990. *Ecology and Management of Aquatic-Terrestrial Ecotones*. UNESCO, Pairs & Parthenon Publ., London.

Needham, J.G., and Lloyd, J.T. 1937. *Life in Inland Waters*. Comstock Pub. Co., Ithaca, New York.

Nelson, J.S. 1994. *Fishes of the World*, 3rd ed. John Wiley & Sons.

Nemerov, N.L., ed. 1974. *Scientific Stream Pollution Analysis*. McGraw-Hill, New York.

Nilsson, A., ed. 1996. *Aquatic Insects of North Europa. A Taxonomic Handbook*. 2 vol. Apollo Books, Stenstrup.

Nordon, M. 1991–1992. *Histoire de l'Hydraulique*, Vol. I: *Les Origines et le Monde Antique*. Vol. II: *L'Eau Démontrée. Du Moyen-Age à Nos Jours*. Masson, Paris.

Nriagu, J.O., ed. 1963. *Aquatic Toxicology*. John Wiley & Sons, Chichester.

Oglosby, R.T., Carlson, C.A., and McLann, J.A., ed. 1972. *River Ecology and Man*. Academic Press, New York.

Patrick, R. 1994–1996. *Rivers of the United States*, 3 vol. John Wiley & Sons, Chichester.

Pennak, R.W. 1978. *Invertebrates of the United States*, 2nd ed, John Wiley & Sons, Chichester.

Pennak, R.W. 1989. *Freshwater Invertebrates of the United States. Protozoa to Mollusca,* 3rd ed., John Wiley & Sons, Chichester.

Person, G., and Jansson, M., ed. 1989. *Developments in Hydrobiology, 48. Phosphorus in Freshwater Ecosystems*. Kluwer Academic Publ., Dordrecht.

Pesson, P., ed. 1980. *La Pollution des Eaux Continentales*. Gauthier-Villars, Paris.

Petts, G.E. 1984. *Impounded Rivers. Perpectives for Ecological Management*. John Wiley & Sons. Chichester.

Petts, G.E., and Calow, P., ed. 1996. *Rivers Flows and Channel Form. Diversity and Dynamics*. Blackwell Sciences, Oxford.

Petts, G.E., and Calow, P., ed. 1996. *River Biota: Diversity and Dynamics*. Blackwell Sciences, Oxford.

Petts, G.E., and Calow, P., ed. 1996. *River Restoration*. Blackwell Sciences, Oxford.

Petts, G.E., Moller, H., and Roux, A.L. 1989. *Historical Change of Large Alluvial Rivers: Western Europe*. John Wiley & Sons, Chichester.

Phillips, D.J.M. 1980. *Quantitative Aquatic Biological Indicators*. Elsevier, Amsterdam.

Pourriot, R., Meybeck, M., and Champ, P., ed. 1995. *Limnologie Générale*. Masson, Paris.

Prach, K., Jemk, J., and Large A.R.G., ed. 1996. *Floodplain Ecology and Management. The Luznice River in the Trebon Biosphere Reserve, Central Europe*. SPB Academic Publ., Amsterdam.

Rai, L.C., Gaur, J.P., and Soeder, C.J., ed. 1994. *Algae and Water Pollution*. Advance in Limnology, no. 42. Schweizerbat'sche Verlag, Stuttgart.

Ramade, F. 1992. *Précis d'Écotoxicologie*. Masson, Paris.

Ramade, F. 1998. *Dictionnaire Encyclopédique des Sciences de l'Eau*. Edisciences, Paris.

Rand, G.M., and Petrocelli, S.R. 1985. *Fundamentals of Aquatic Toxicology*. Hem. Publ., Corp., Washington.

Resh, V.H., and Rosenberg, D.M., ed. 1984. *The Ecology of Aquatic Insects*. Praeger, New York.

Reid, J.K., and Wood, R.D. 1976. *Ecology of Inland Waters and Estuaries*. Van Nostrand, New York.

Richard K.S., ed. 1987. *River Channels: Environment and Process*. Blackwell Sciences, Oxford.

Rhodes, D.D., and Williams, G.P., ed. 1979. *Adjustments of the Fluvial System*. Kendal Hunt Publ. Co., Dubuque, Iowa.

Ryding, S.O., and Rast, W., ed. 1994. *Le Contrôle de l'Eutrophisation des Lacs et des Réservoirs*. Masson, Paris.

Rzoska, J., ed. 1976. *The Nile. Biology of an Ancient River*. W. Junk, The Hague.

Saila, S.A., ed. 1975. *Fisheries and Energy Production. A Symposium*. Lexington, Massachusetts.

Schiemer, F., and Boland, K.T., ed. 1996. *Perspectives in Tropical Limnology*. SPB Academic Publ., Amsterdam.

Schiemer, F., Zalewski, M., and Thorpe, J.E., ed. 1995. *The Importance of Aquatic-Terrestrial Ecotones for Freshwater Fish*. Kluwer Acad. Publ., Dordrecht.

Schönborn, W. 1992. *Fliessgewässerbiologie*. G. Fischer Verl., Stuttgart.

Schubert, L.E., ed. 1984. *Algae as Ecological Indicators*. Academic Press, London.

Schumm, S.A. 1977. *The Fluvial System*. Wiley Interscience, New York.

Sculthorpe, C.D. 1967. *The Biology of Aquatic Vascular Plants*. Edward Arnold, London.

Schwabe, A. 1987. *Fluss- und Bachbegleitende Pflanzengesellschaften und Vegetationskomplexe in Schwarzwald*. E. Schweizerbart'sche, Stuttgart.

Schwoerbel, J. 1996. *Methoden der Hydrobiologie*. Kosmos. Franckh'sche Verlag, Stuttgart.

Schwoerbel, J. 1974. *Einführung in die Limnologie*. G. Fischer, Stuttgart.

Schwoerbel, J. 1987. *Handbook of Limnology*. J. Wiley & Sons, Chichester.

Sigg, L., Stumm, W., and Behre, P. 1992. *Chimie des Milieux Aquatiques*. Masson, Paris.

Simon, M., Güde, H., Weisse, T., ed. 1996. *Aquatic Microbial Ecology*. Advances in Limnology, 48. E. Schweizerbart'sche, Stuttgart.

Sing, V.P. 1995. *Environmental Hydrology*, Kluwer Acad. Publ., Dordrecht.

Skinner, F.A., and Shewan, J.M., ed. 1977. *Aquatic Microbiology*. Academic Press, London.

Sladecek, V. 1973. *System of Water Quality from the Biological Point of View*. Advances in Limnology, Schweizerbart'sche Verlag, Stuttgart

Sladecek, V., ed. 1978. *Symposium in Saprobiology*. Advances in Limnology, 9. E. Schweizerbart'sche, Stuttgart.

Smart, M.M., Lubinski, K.S., and Snick, R.A. 1986. *Ecological Perspectives of the Upper Mississipi River*. W. Junk, The Hague.

Sorokin, Y.I. 1999. *Aquatic Microbial Ecology*. Backhuys Publ., Leiden.

Spillman, C.J. 1961. *Poissons d'Eau Douce*. Faune de France, no. 65. P. Lechevalier, Paris.

Straskraba, M., ed. 1993. *Comparative Reservior Limnology and Water Quality Management*. Kluwer Acad. Publ., Dordrecht.

Sutcliffe, D.W., ed. 1994. *Water Quality and Stress Indicators in Marine and Freshwater Systems: Linking Levels and Organization*. Freshwater Biological Association, Great Britain.

Sutcliffe, D.W., ed. 1996. *The Ecology of Large Rivers*. Suppl. Arch. Hydrobiol. E. Schweizerbart'sche, Stuttgart.

Symoens, J.J., Hooper, S.S., and Compere, P., ed. 1982. *Studies on Aquatic Vascular Plants*. Bot. Roy. Soc., Belgium.

Tachet, H., Bournaud, M., and Richoux, P., 1987. *Introduction à l'Étude des Macroinvertébrés des Eaux Douces*. A.F.L. and Université Lyon I.

Tardy, Y. 1986. *Le Cycle de l'Eau*. Masson, Paris.

Thibault, M., and Billard, R., ed. 1987. *Restauration des Rivières à Saumon*. INRA, Paris.

Thorne, C.R., Bathurst, J.C., and Hey, R.D., ed. 1987. *Sediment Transport in Gravel-bed Rivers*. John Wiley & Sons, Chichester.

Uhlmann, D. 1975. *Hydrobiologie*. G. Fisher, Iéna.

Vaillant, J.R. 1973. *Protection de la Qualité des Eaux et Maîtrise de la Pollution*. Eyrolles, Paris.

Vibert, R., and Lagler, K.F. 1961. *Pêches Continentales, Biologie et Aménagement*. Dunod, Paris.

Vincent, W.F., and Ellis, J. ed. 1988. *High Latitude Limnology*. Development in Hydrobiology. Kluwer Acad. Publ., Dordrecht.

Vogel S. 1996. *Life in Moving Fluids. The Physical Biology of Flow*. Princeton Univ. Press, New York.

Ward, J.W. 1992. *Aquatic Insect Ecology*, 2 vol. John Wiley & Sons, Chichester.

Ward, J.V., and Standford, ed. 1979. *The Ecology of Regulated Streams*. Plenum Press, New York.

Welcome, R.L. 1979. *Fisheries Ecology of Floodplain Rivers*. Longmans, London.

Wetzel, R.G. 1983. *Limnology*, 2nd ed. Saunders College Publ., Philadelphia.

Wetzel, R.G., ed. 1983. *Periphyton of Freshwater Ecosystems*. Developments in Hydrobiology, 17. W. Junk, The Hague.

Whitton, B.A., ed. 1975. *River Ecology*. Studies in Ecology, vol. 2. Blackwell Scient. Publ., London.

Whitton, B.A., Rott, E., and Friedrich, G., ed. 1991. *Use of Algae for Monitoring Rivers*. Inst. Botanik, Innsbrück.

Williams, W.D., and Sladecekova, A., ed. 1996. *Water as Limiting Resources: Conservation and Management*. Proc. Sil.E. Schweizerbart'sche, Stuttgart.

Zajak, Z., and Hillbricht, A., ed. 1970. *Productivity Problems of Freshwaters*. UNESCO IBP Symposium, Warsaw.

Index

Acclimatization (of fishes) 74
Acidophile species 96
Adaptaion of current 27
Allen paradox 48
Allochthonous materials 101, 108
Alluvial forest 14, 133
Alluvial plain 41, 46, 102, 105, 127, 131, 132, 133, 134, 137
Altitudinal 78, 84, 87, 88, 89, 90, 92
Ammonia 8, 133, 163, 168, 172, 173, 174, 186, 198
Amplifying factors 198
Anadromous migration 67, 72
Analysis
- canonical 18
- comparative analysis of communities 184

Arrhenius law 78
Arsenic 172
Assessment
- of drift 33
- of material transport 9

Aprons 110
Assimilation capacity 164, 166, 169
Autochthonous plant production 103, 151
Autotrophy 105, 196
Autotrophy index 183

Benthos 32, 33, 35, 36, 37, 39, 40, 48, 109, 127, 134, 142, 146, 150, 162, 165, 185, 195, 197
Bioaccumulation 175, 176
Bioassays 183
Biodegradation 22, 23, 24, 52, 101, 102, 105, 155, 157, 160, 161, 162, 163, 164, 167, 168, 169, 170, 172, 175, 196
Biogeography
- of stygobious fauna 46
- of fishes 78

Biogenic capacity 112
Biological cover 90, 103, 104, 107, 108, 109, 111, 122, 164
Biological cycle 77, 84
Biological indicators 121, 185, 186, 189, 192
Biosynthesis 23, 52, 101, 170, 196
Bivoltin species 88
Bloom 64, 66, 155, 158
Boundary layer 27, 29, 51, 164
Braided beds 13
Browsing (invertebrates) 65, 66, 109, 152, 160, 168, 196
Bryophytes 21, 22, 27, 49, 50, 51, 52, 94, 95, 96, 122, 123, 124, 125, 126, 142, 176, 184, 196

Cadmium 172, 175, 176, 177, 180
Calcicolous species 96
Calcium 7, 95, 96, 97, 113, 114, 115, 174
Canals 59, 65, 73, 75, 126, 139, 150, 189, 198
Carbon
- algal 104
- detritic 105
- particulate 167

Carbonic anhydride 6, 8, 96, 97, 98, 151, 162
Cascades 113, 120, 124
Catadromous migration 67, 72
Chlorophyll pigments (algae) 93

Chromium 175
Cobalt 175
Competence of current 4
Consumers 23, 59, 105, 106, 108, 116, 176, 186
Correlations (between parameters) 18
Crenal 122, 123, 125, 126, 129, 130, 132, 189
Crenon 122, 123, 142, 189
Cumulative effect 175

Degree-days 17, 18, 82, 83, 84, 85, 86, 124, 199
Detergents 8, 151, 159, 174
Detritivores invertebrates) 107, 108, 109, 111, 163, 165
Development
  multiple 139
  in the rhithral 140
Development constant 82, 83, 84, 85, 86, 88
Development, duration of 82
Diapause 84, 85, 86, 87, 89, 90, 91, 92
Dilacerators (invertebrates) 103, 108, 109
Dissolved oxygen 13, 45, 93, 97, 98, 99, 123, 128, 156, 157, 161, 163, 164, 165, 166, 169, 170, 171, 172, 173, 198
Distribution of organisms
  altitudinal 88
  of fishes 69, 72
  of hyporheic fauna 43
Diversity 21, 43, 73, 92, 114, 123, 124, 147, 150, 177, 185, 192
Drift
  inert 30
  active 34

Ecoregions 193
Ecosystems
  running water, lotic 24
Ecotone 39
Electrolytes 6, 7, 8, 9, 95, 96, 97
Energy flows 101, 115
Epipotamal 121, 129
Epirhithral 121, 124, 125, 126, 129, 130
Equilibrium 4, 5, 8, 12, 13, 18, 36, 162, 164, 170, 173, 199
Erosion 1, 2, 4, 5, 7, 9, 12, 13, 18, 24, 30, 35, 103, 120, 123, 125, 131, 140, 152, 166, 195, 198
Eucrenal 121, 122
Eucrenon 122, 123
Euphotic zone 94
Eurythermic species 78, 80
Eutrophy 151
Eutrophication 8, 65, 104, 149, 151, 152, 153, 155, 157, 159, 160, 161, 164, 169, 170, 174, 184, 186, 187, 197, 198, 199
Evaluation of pollution
  bioassays 183
  biochemical methods 183
  biocoenotic methods 184
  comparative analysis of
    communities 184
  ecotoxicological methods 183
Extendens biotic index 187

Factorial analysis of correspondences 19, 118
Falls 1, 15, 23, 49, 84, 124, 132, 158, 163, 168, 173, 174, 177, 178, 196, 120
Field (concept of) 195
Filter feeders (invertebrates) 165
Fish 21, 22, 27, 56, 57, 59, 67, 68, 69, 71, 72, 73, 74, 75, 80, 81, 83, 84, 98, 99, 111, 112, 114, 115, 121, 129, 134, 143, 144, 145, 147, 148, 150, 161, 170, 172, 173, 174, 175, 176, 183, 184, 187, 195
Flats 113, 120, 125, 126, 127, 130
Flight periods 87, 88
Flood plain 102, 105, 133, 134, 195, 196, 197
Floods 14, 33, 37, 39, 61, 75, 131, 132, 134, 139, 148, 149, 156, 196, 197
Flow
  in plains 13

reserve 149
regulated 145
of water 12, 15
Fluoride 175
Fluvial continuum 105, 109, 116, 129, 130, 134, 137
Food regime 130, 176
Food web 108
Force of current 2
Fungicides 172, 174

Granulometry (of substrate) 5, 41, 43, 71, 118, 120, 164
Grinding (invertebrates) 107

Heavy metals 175, 176, 181, 184, 198
Heliophile species 94
Helocrene sources 122
Herbarium 56
Herbicides 175
Heterotrophy 105, 170, 196
Hjülstrom curves 6
Horton (order of drainage) 10
Hydrogen sulphide 172, 173, 186, 198
Hydrographic Network 11, 125
Hydrosystem 14
Hygropetric surface 49
Hypocrenal 121, 122, 123
Hypocrenon 123
Hypomycetes (fungi) 102, 103
Hypopotamal 129
Hyporheic environment 39, 40, 41, 42, 43, 45, 46, 48, 59, 98, 99, 102, 108, 117, 123, 125, 126, 129, 136, 147, 150, 196
Hyporhithral 121, 124, 125, 126, 127, 129

Impact of development 139
Indefinite survival (limit of) 80, 173, 174
Index
biotic (Tuffery and Verneaux) 190
chironomid 187
diversity 185
diatomic 186
macrophyte 187
Oligochaeta 187
potential biological quality 189
similarity 185
standard global biological 189
Insecticides 172, 174
Interstitial environment 39, 41, 46, 47, 48, 51, 108

Lead 3, 17, 51, 72, 77, 98, 99, 152, 155, 156, 162, 169, 170, 175, 176, 177, 198
Lethal temperatures 80
Light 23, 31, 41, 49, 55, 61, 63, 64, 65, 83, 93, 94, 95, 96, 97, 103, 109, 115, 125, 127, 152, 157, 159, 169, 170, 196, 198, 199
Limiting factors 18, 115, 116, 155, 157
Limnocrene sources 22, 122, 123
Litter 98, 101, 102, 103, 125
Littoral zone (mobile) 22
Load limit 4
Lone 13, 121
Low water 13, 15, 17, 18, 33, 51, 52, 56, 65, 99, 102, 104, 105, 109, 111, 116, 120, 125, 126, 127, 128, 132, 133, 134, 135, 137, 140, 141, 142, 149, 157, 166, 195, 196, 197

Macrophyte 22, 37, 49, 54, 94, 96, 101, 103, 109, 122, 124, 142, 149, 151, 152, 153, 156, 160, 174, 176, 184, 185, 186, 187, 192, 196, 198
Manning equation 2
Margalef formula 36, 51, 87
Meanders 13, 14, 120, 121, 125, 127
Mediterranean 15, 16, 17, 35, 43, 46, 47, 73, 82, 134, 137
Mercury 172, 175
Mesosaprobe zone 180
Mesotrophy 151
Metabolism 40, 77, 78, 79, 80, 81, 84, 98, 111, 115
Metapotamal 121, 129

Metarhithral 121, 124, 125, 126, 129, 130, 164
Microhabitats (method of) 37, 46, 56, 104, 109, 113, 114, 115, 116, 118, 119, 122, 123, 124, 125, 127, 129, 130, 142, 147, 150, 183, 193, 195
Microphagous species 106
Micropollutants
cumulative effect 175
transfer along food web 175
Migration (fishes) 36, 43, 87, 137
Modelling of phytoplankton development 62
Monovoltin species 92, 110
Muds 120, 125, 130

Necromass 23
Nitrates 8, 52, 133, 151, 153, 154, 156, 162, 168, 186
Nitrogen 8, 115, 133, 151, 152, 155, 159, 160, 161, 162, 163, 168, 170, 186
Nodas 130

Oligosaprobe zone 162, 186
Oligotrophy 151
Omnivores
invertebrates 106
fishes 109
Organochlorine 175
Organophosphates 174

Periphyton 83, 151
Pesticides 184
Phanerogams 52, 53, 54, 56, 126
Phenol 172
Phosphates 8, 52, 133, 153, 154, 156, 159, 160, 162
Phosphorus 8, 115, 151, 152, 153, 155, 156, 159, 160, 161, 162, 168, 169, 170, 186
Photoperiodism 95
Phytobenthos 103, 104, 152, 153
Phytophagous species 109
Phytoplankton 103, 104, 115, 116, 151, 152, 153, 154, 155, 156, 157, 158, 159, 160, 167, 169

Plankton 18, 115, 127, 134
Plant production(autochthonous) 103
Plant score 186
Pluvial oceanic 17, 104, 155
Pollution
chemical 174
mechanical 171
multiple 178
salinity 174
toxic 171
toxic organic 160
trophic 151, 161, 169
Pollution resistance 181
Pollution sensitivity 182, 184
Polyphosphates 8, 151, 174
Polysaprobe zone 162, 180, 186
Polyvoltin species 79, 85, 92, 110, 127, 135
Porosity (of substrate) 43
Potamal 12, 43, 48, 78, 102, 111, 114, 122, 127, 128, 129, 130, 131, 136, 152, 155, 164, 165, 166
Potamon 122, 127, 135
Predators (invertebrates) 107, 108, 109, 110
Production/biomass ratio (P/B) 110
Production/respiration ratio (P/R) 105
Profile
equilibrium 4
longitudinal 5
Pteridophytes 49

Q_{10} 78
Quiescence 84, 86, 87, 89, 91

Rapids 120, 124, 125
Receiving capacity (fishes) 112
Redistribution of materials 18
Regulatory factors 153, 170
Resurgence (sources) 122
Reynolds numbers 3
Rheocrene sources 122, 123
Rheophile species 133
Rhithral 12, 43, 50, 78, 122, 123, 124, 152, 159, 164, 166, 170
Rhithron 122, 124, 135, 145
Rhizoids 49

Salts, dissolved, 93, 95
Saprobe zone 166
Saprobe system 186
Sciaphile species 94
Scraping of substrate (invertebrates) 107
Self-purification
  in the rhithral 164
  in the potamol 165
  in the Seine 166
Semi-voltin species 79
Sluices 140, 145
Spermatophytes 51, 94
Spiralling 170
Station 18, 19, 20, 167
Stenothermic species 79, 135
Strahler (order of drainage) 11
Stygobious species 41, 43, 45
Stygophilous species 40
Stygoxenous species 40
Sub-flow 13, 39, 44, 87
Succession
  spatio-temporal 91
  upstream-downstream 65, 81
Swimming (of fishes) 67

Temperature
  acclimatization 80
  lethal 76, 80
  maximum activity 76
  threshold 76
Thermal sum 17
Thigmotactism 40
Threshold 77, 80, 82, 83, 84, 88, 176, 192
Toxicity tests 175, 183
Transfer (canal and alluvial plain) 132
Transit time 9, 17, 18, 22, 25, 59, 61, 62, 64, 65, 66, 104, 115, 127, 142, 146, 149, 152, 154, 155, 156, 157, 158, 159, 160, 161, 162, 164, 166, 168, 169, 170, 195, 196, 197, 198, 199
Transport of materials
  by carrying 6
  in solution 6
  in suspension 6
Trent biotic index 187, 189
Tress beds 131, 133
Trout zone 69, 70, 71, 72, 74, 123, 124, 129
Turbulence 3, 4, 5, 6, 12, 13, 65, 66, 94, 98, 120, 123, 127, 164, 192

Useful weighted area 134

Vollenweider model 151

Water cycle 1, 2
Water regime
  Mediterranean 17, 134
  multiple 17
  nival 15
  pluvial oceanic 15
Waters
  dead 13
  of first category 70
  of second category 71
Watershed 11, 12, 13, 15, 16, 17, 18, 19, 20, 24, 55, 101, 102, 103, 121, 124, 125, 130, 131, 133, 151, 161, 162, 164, 195, 196, 197

Xenosaprobe zone 186

Zinc 171, 172, 175, 176, 177, 180
Zonation, ecological
  upstream-downstream 116, 121
  Huet 69, 121, 129
  Illies and Botosaneanu 115, 121, 129, 192
Zones 2, 9, 12, 13, 16, 17, 22, 44, 48, 49, 50, 61, 69, 70, 71, 72, 74, 79, 80, 87, 90, 99, 109, 111, 114, 120, 121, 124, 125, 126, 127, 129, 130, 133, 136, 145, 149, 162, 163, 166, 189, 196
Zooplankton 23, 59, 62, 65, 66, 152, 156, 160